# BLAST OFF to 3rd Grade Math

## 2nd to 3rd Grade Transition Math Workbook

This book belongs to:

______________________________

# GET THE ANSWER KEY

The answer key for this book is available as a free PDF download.

SCAN HERE



This product is not sponsored or endorsed by the Common Core State Standards Initiative of the National Governors Association Center for Best Practices and the Council of Chief State School Officers.

Photos of coins and currency are courtesy of the United States Mint and the United States Treasury Bureau of Engraving and Printing.

Printed in the United States of America
ISBN: 978-1-64824-039-3
EDventure Learning LLC
5601 State Route 31 #1296
Clay, NY 13039

www.edventurelearning.com

Email us at hello@edventurelearning.com
Follow us on social media @edventurelearning

2 - 3

This book is part of the **Bridge Builder Learning Series**, designed to help bridge the gap between grade levels. The content in this book is designed to review math from 2nd grade and set students up for success in 3rd grade.

## CONTENTS

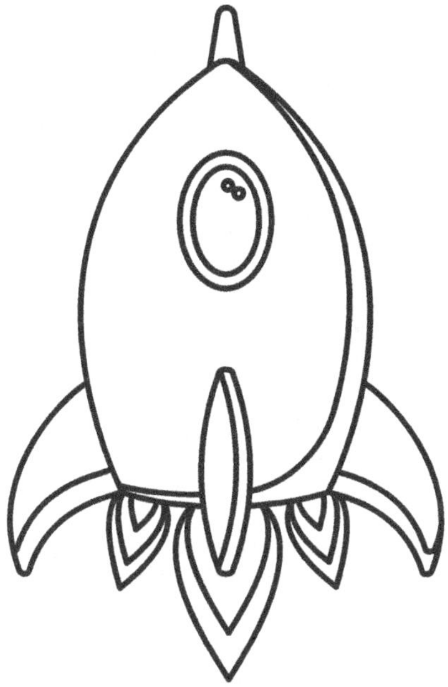

# Numbers and Counting

# 6 Place Value

Use the number of tens and ones to find the total number.

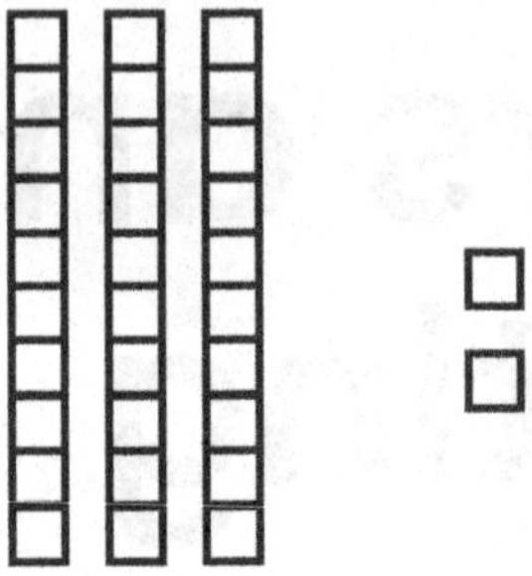

3 tens + 2 ones = 32

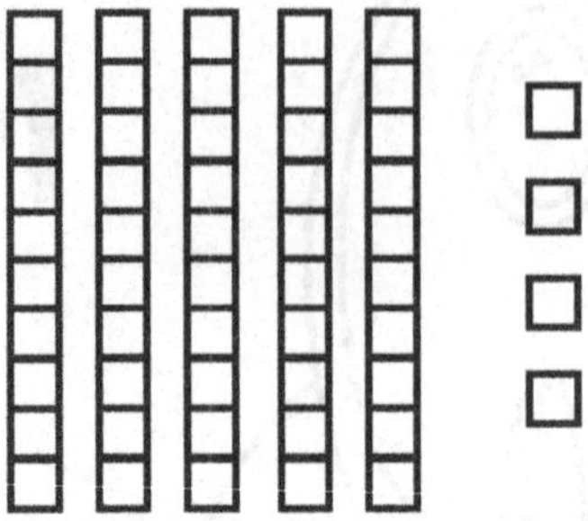

____ tens + ____ ones = ____

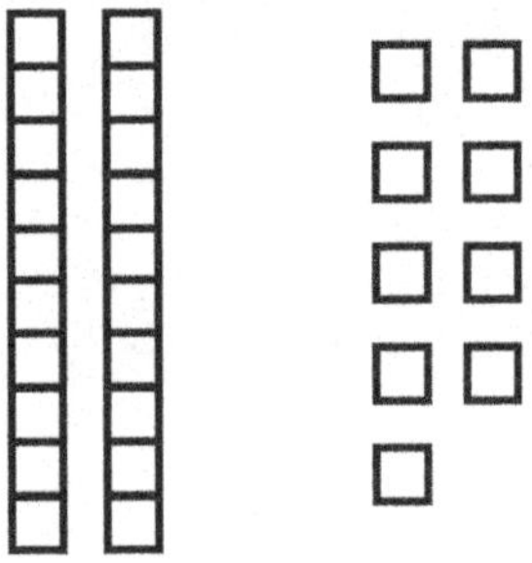

____ tens + ____ ones = ____

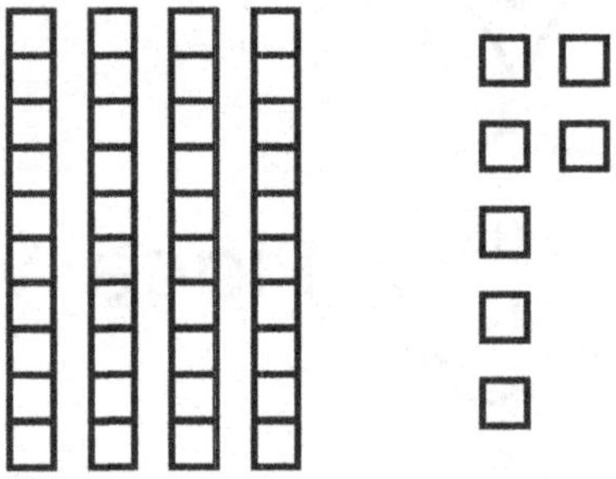

____ tens + ____ ones = ____

Fill in the missing number in each equation to build a two-digit number.

(EX) 4 tens + 6 ones = 46

(1) 7 tens + ____ ones = 72

(2) ____ tens + 1 one = 91

(3) 2 tens + 6 ones = ____

(4) 30 + ____ = 38

(5) 70 + 5 = ____

(6) 60 + ____ = 63

(7) ____ + 4 = 84

# Place Value



Write the number represented by the tens and ones shown in each picture.

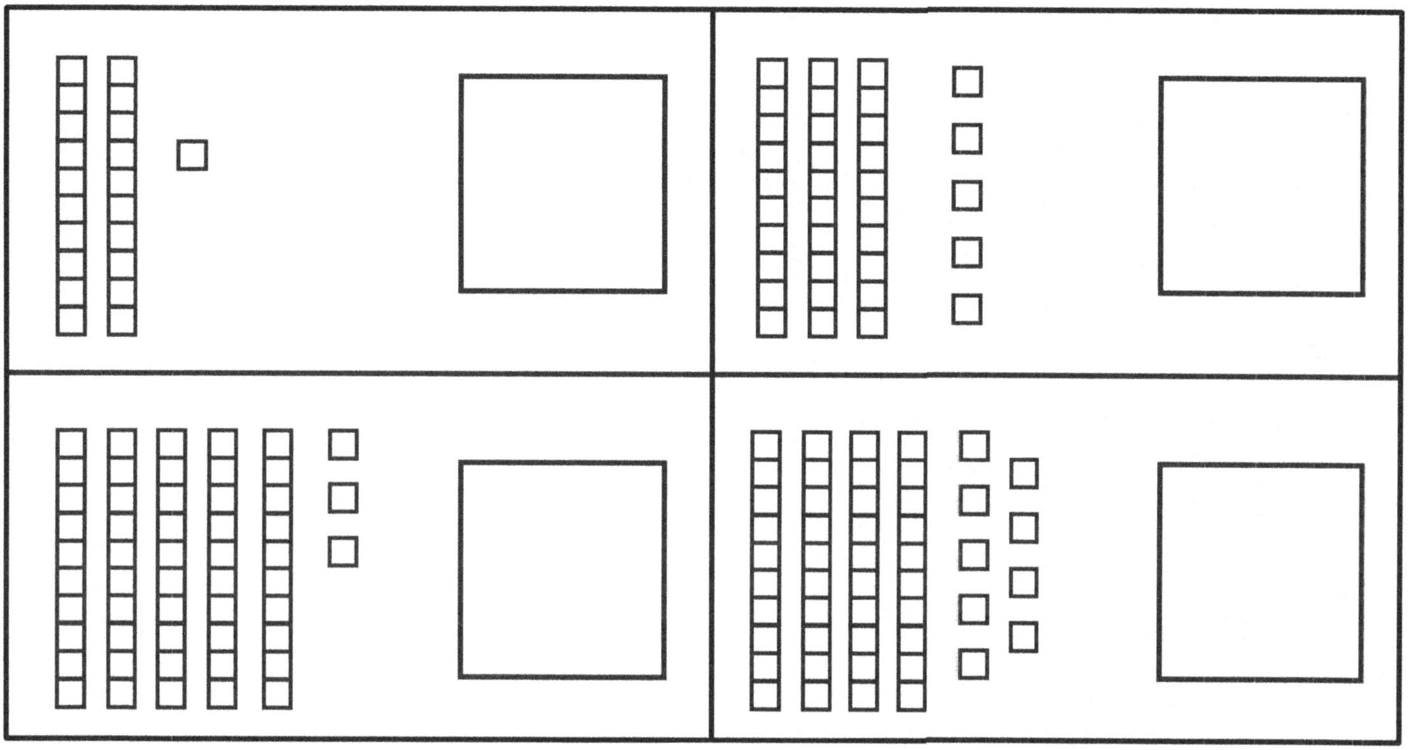

A **quick picture** uses a stick to represent each ten and a circle to represent each one. Make a quick picture to show each number.

| | |
|---|---|
| 51 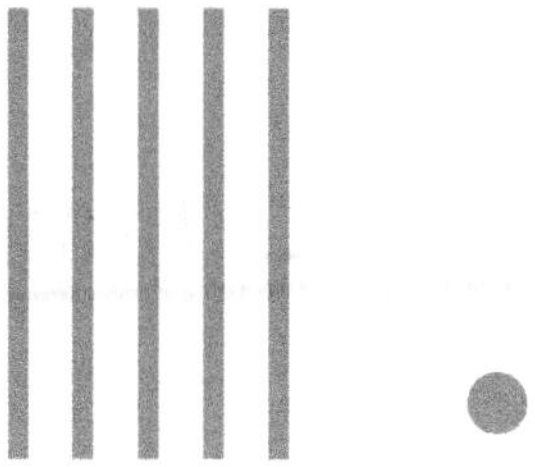 | 97 |
| 18 | 72 |

# 8 Place Value

Circle groups of tens to make hundreds. Then, fill in the blanks to write the total three ways.

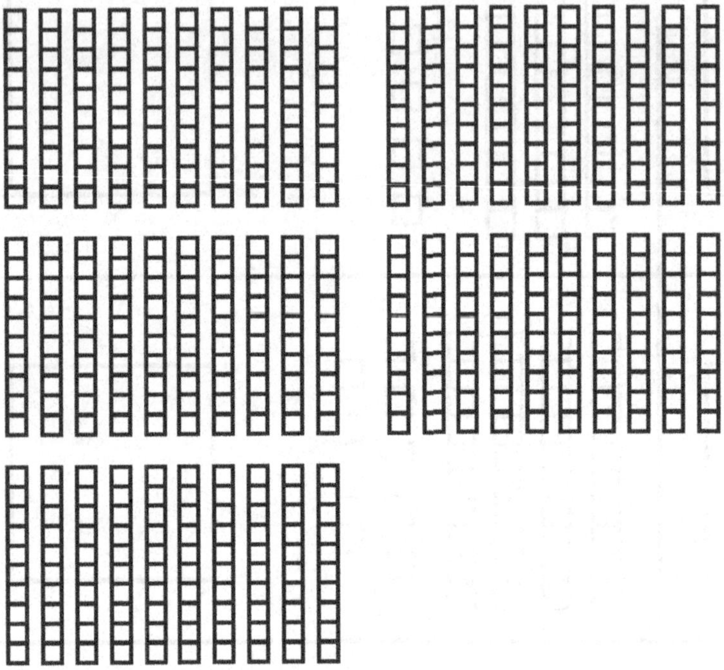

_______ hundreds =

_______ tens =

_______ ones

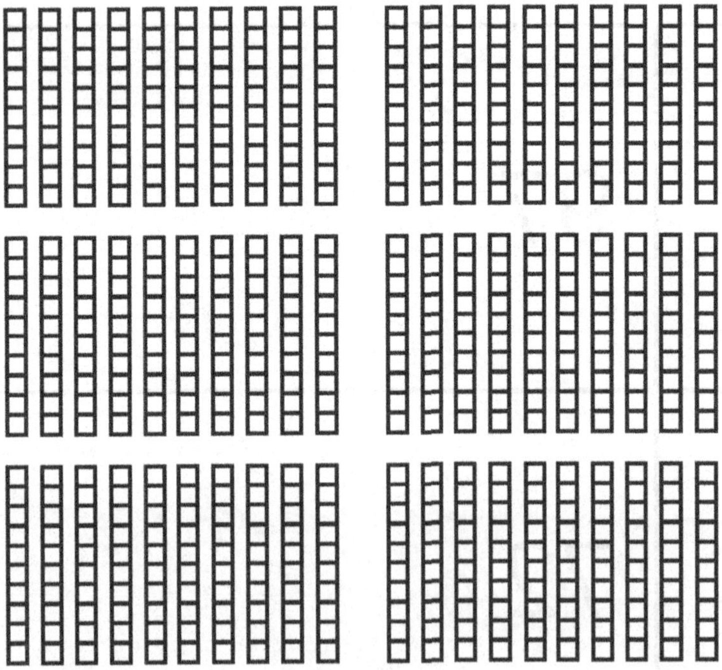

_______ hundreds =

_______ tens =

_______ ones

Draw a quick picture to show the problem. Circle tens to make hundreds and write the answer on the line.

| | |
|---|---|
| Each student in the class has a box of 10 crayons. There are 20 students in the class. How many crayons do they have all together?<br><br>______ crayons | |
| Reese puts crackers in bags. She fills 30 bags. Each bag has 10 crackers in it. How many crackers does Reese put in bags?<br><br>______ crackers | |
| Each book in a series has 40 pages. There are 10 books in the series. How many pages are in the books all together?<br><br>______ pages | |

# 10 Place Value

In each number shown, color the digit in the hundreds place RED, the digit in the tens place **BLUE**, and the digit in the ones place **GREEN**. Then, color in the number cubes below each digit with the same colors.

1 3 6

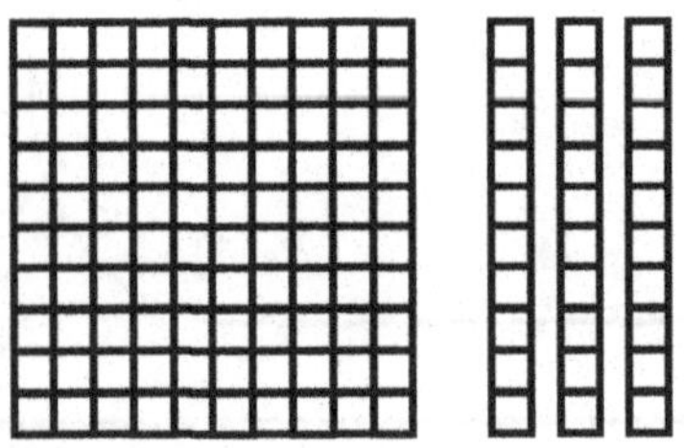

2 1 8

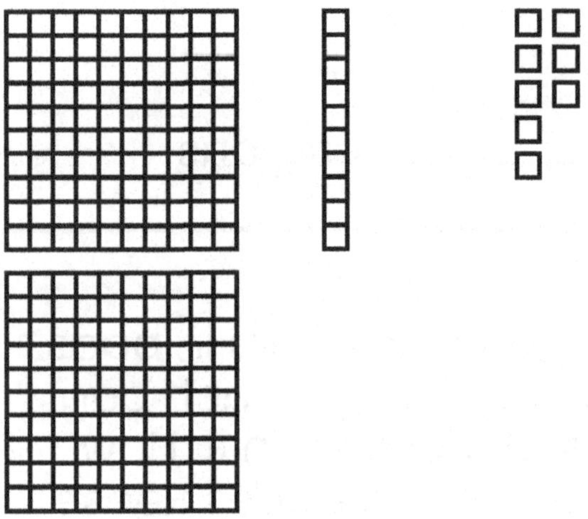

4 7 5

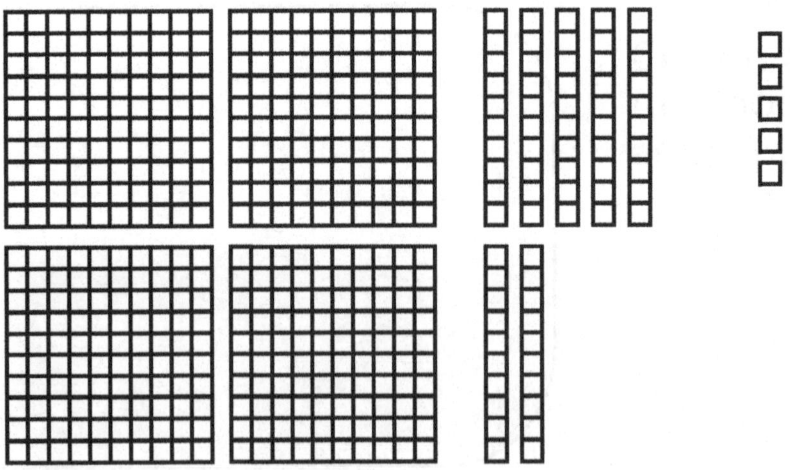

For each number, write the digits in the correct column in the place value chart.

562

| Hundreds | Tens | Ones |
|---|---|---|
| 5 | 6 | 2 |

289

| Hundreds | Tens | Ones |
|---|---|---|
| | | |

107

| Hundreds | Tens | Ones |
|---|---|---|
| | | |

650

| Hundreds | Tens | Ones |
|---|---|---|
| | | |

784

| Hundreds | Tens | Ones |
|---|---|---|
| | | |

422

| Hundreds | Tens | Ones |
|---|---|---|
| | | |

91

| Hundreds | Tens | Ones |
|---|---|---|
| | | |

300

| Hundreds | Tens | Ones |
|---|---|---|
| | | |

# 12 Place Value

Write how many hundreds, tens, and ones are shown in each picture. Then, write the number.

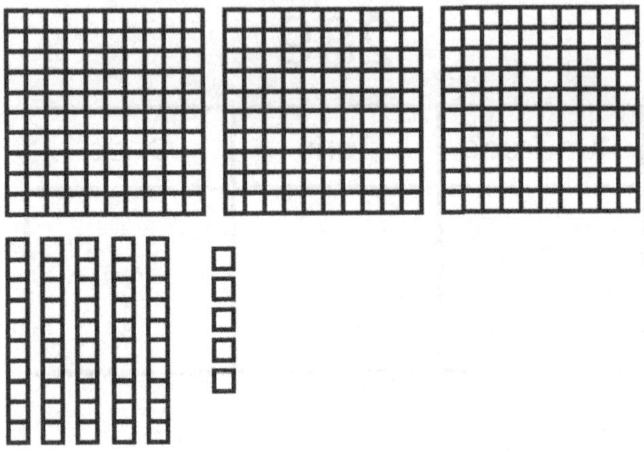

| Hundreds | Tens | Ones |
|---|---|---|
| | | |

____________

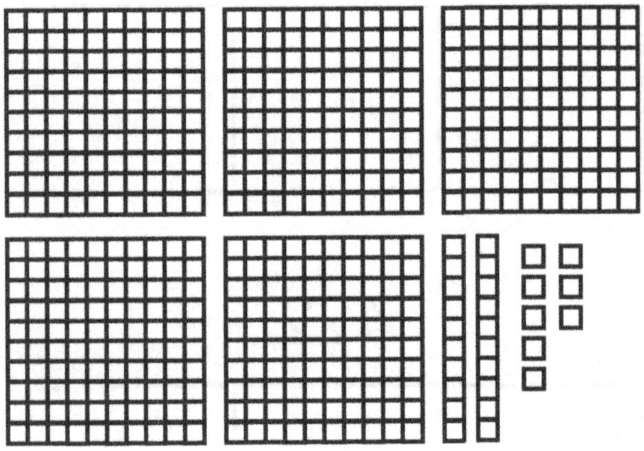

| Hundreds | Tens | Ones |
|---|---|---|
| | | |

____________

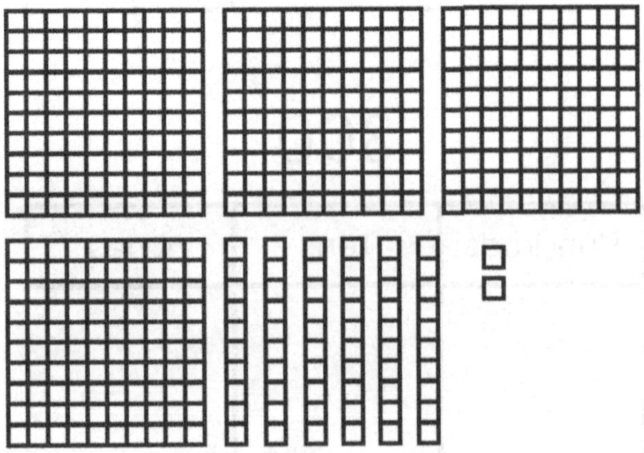

| Hundreds | Tens | Ones |
|---|---|---|
| | | |

____________

# Place Value



Look at the picture and write the number of blocks three ways.

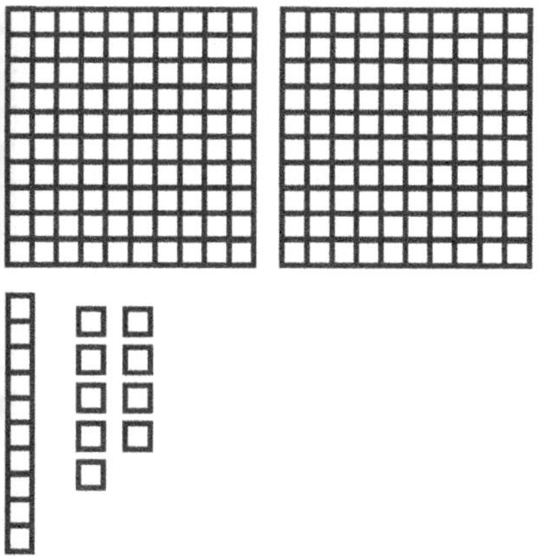

______ hundreds, ______ tens, ______ ones

________ + ________ + ________

__________

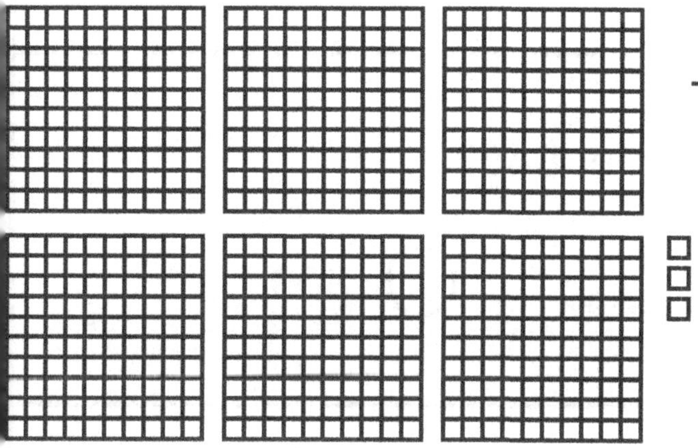

______ hundreds, ______ tens, ______ ones

________ + ________ + ________

__________

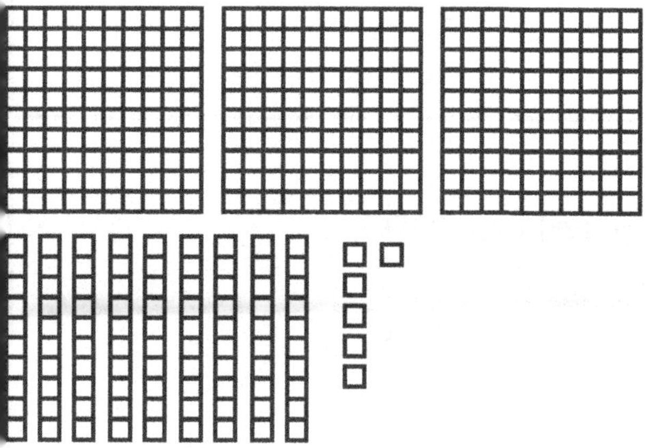

______ hundreds, ______ tens, ______ ones

________ + ________ + ________

__________

# Place Value

Fill the in correct bubble to answer each question.

What does the 4 represent in the number 541?

○ 4 hundreds ○ 4 tens ○ 4 ones

What does the 9 represent in the number 329?

○ 9 ones ○ 9 tens ○ 9 hundreds

What does the 8 represent in the number 845?

○ 8 ○ 80 ○ 800

Which digit in is the tens place in the number 374?

○ 7 ○ 3 ○ 4

Which digit is in the ones place in the number 709?

○ 0 ○ 7 ○ 9

Write the three-digit number that answers each riddle.

| | |
|---|---|
| My hundreds digit is 4.<br>My tens digit has a value of 20.<br>My ones digit is one less than my hundreds digit.<br>What number am I? | |
| My tens digit and hundreds digit are the same.<br>My ones digit is 2.<br>My tens digit and hundreds digit add up to 10.<br>What number am I? | |
| My hundreds digit has a value of 200.<br>My ones digit is the same as my hundreds digit.<br>My tens digit is 7.<br>What number am I? | |
| The value of my tens digit is 0.<br>My ones digit is 8.<br>My hundreds digit is greater than my ones digit.<br>What number am I? | |
| My tens and ones digits add up to my hundreds digit.<br>My ones digit is 2.<br>My tens digit is one more than my ones digit.<br>What number am I? | |

# Representing Numbers

Fill the in correct bubble to answer each question.

Which picture represents 234?

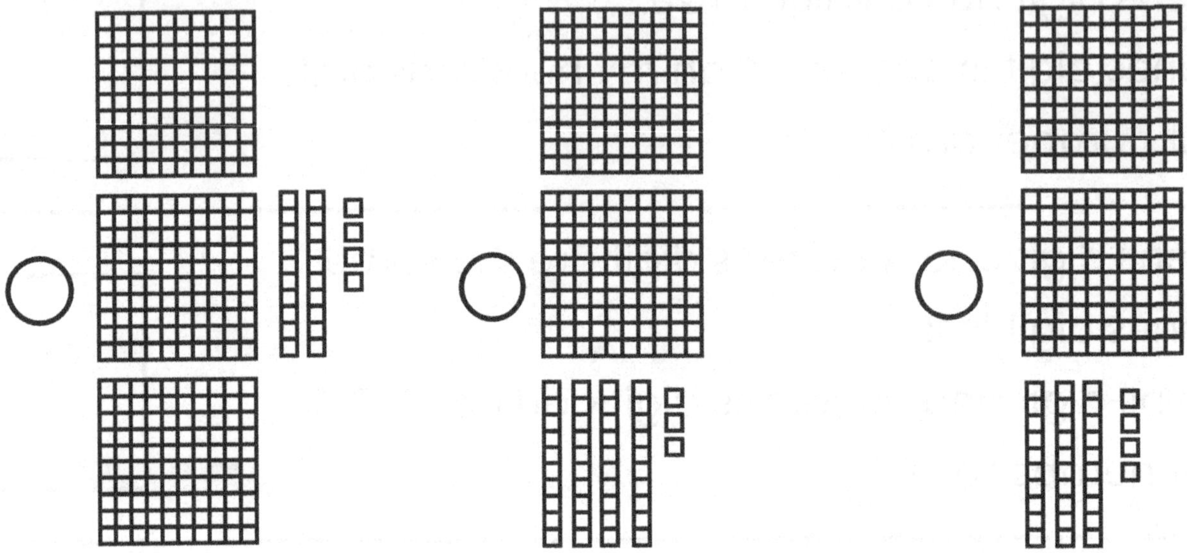

Which picture represents 301?

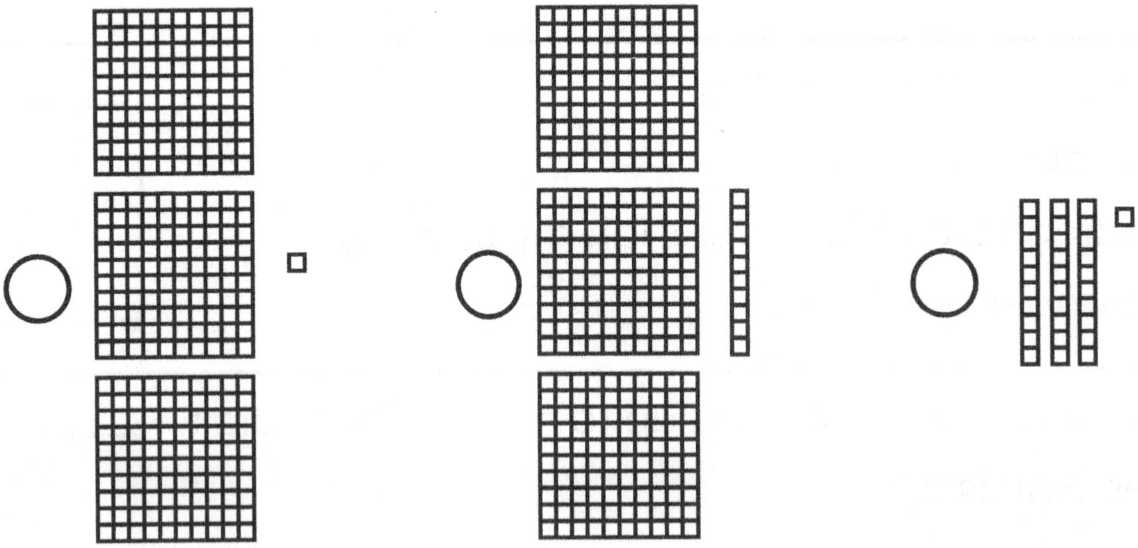

# Representing Numbers



Write each number in expanded form.

1. 287 = _____ + _____ + _____

2. 692 = _____ + _____ + _____

3. 345 = _____ + _____ + _____

4. 463 = _____ + _____ + _____

5. 199 = _____ + _____ + _____

---

Change each expanded form into a number in standard form.

6. 300 + 70 + 8 = _____

7. 200 + 10 + 5 = _____

8. 800 + 20 + 1 = _____

9. 100 + 30 + 7 = _____

10. 500 + 90 + 3 = _____

11. 600 + 40 + 2 = _____

12. 400 + 50 + 9 = _____

13. 700 + 60 + 4 = _____

# 18 Representing Numbers

Write each number in standard form.

1. One hundred thirty-seven ________
2. Three hundred twenty-one ________
3. Six hundred ninety-three ________
4. Nine hundred eighty-two ________
5. Five hundred seventeen ________

---

Write out the name of each number.

6. 845 ____________________
7. 286 ____________________
8. 314 ____________________
9. 592 ____________________
10. 488 ____________________

# Representing Numbers

Complete the diagram about each number.

| NAME | PICTURE |
|---|---|
| | |
| **548** | |
| PLACE VALUES<br>_____ hundreds<br>_____ tens<br>_____ ones | EXPANDED FORM<br>_____ + _____ + _____ |

| NAME | PICTURE |
|---|---|
| | |
| **297** | |
| PLACE VALUES<br>_____ hundreds<br>_____ tens<br>_____ ones | EXPANDED FORM<br>_____ + _____ + _____ |

| NAME | PICTURE |
|---|---|
| | |
| **163** | |
| PLACE VALUES<br>_____ hundreds<br>_____ tens<br>_____ ones | EXPANDED FORM<br>_____ + _____ + _____ |

# 20 Representing Numbers

Find the dot on each number line. Write its value in the box.

1

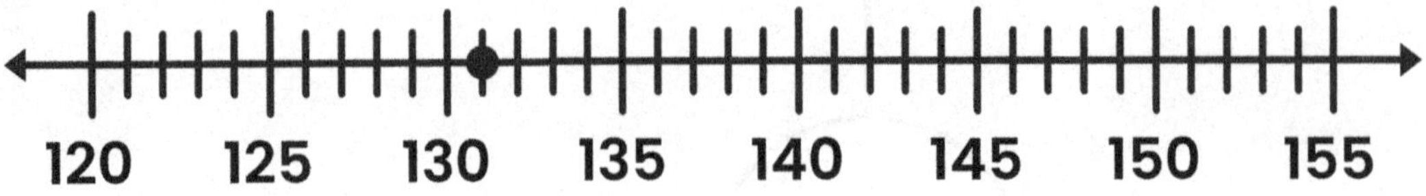


2

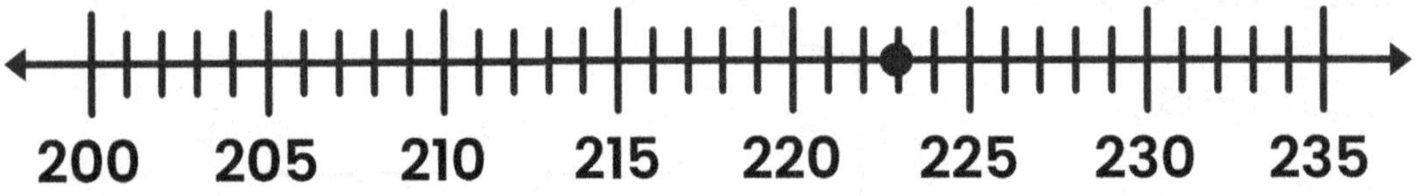


3

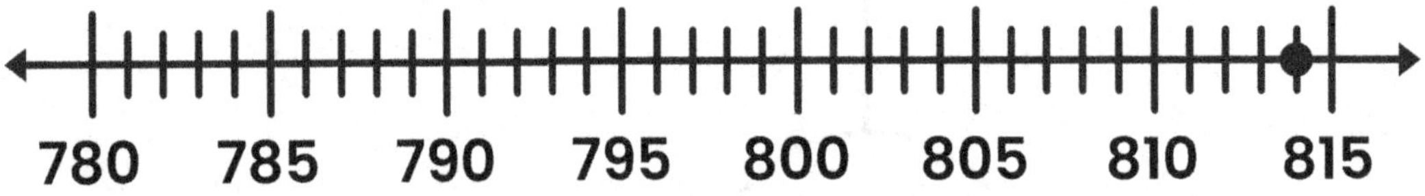


4

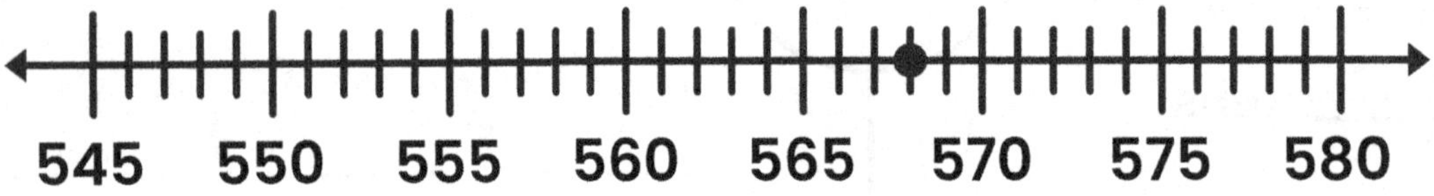


5

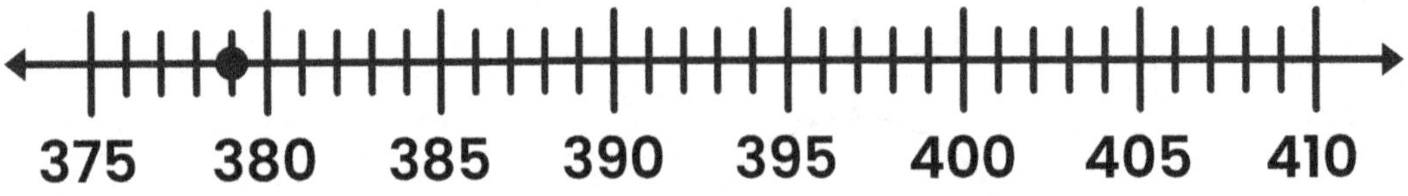


6

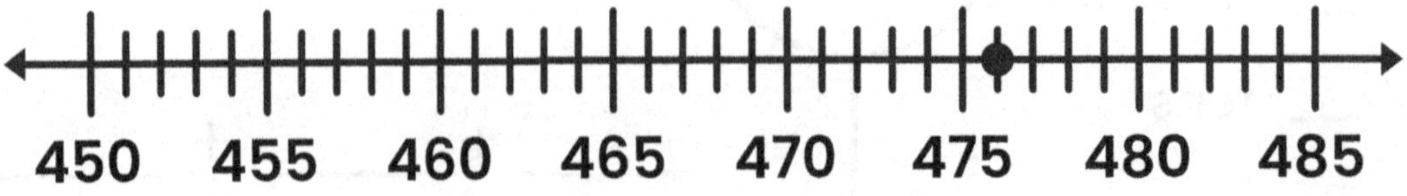

Mark each number with a dot on the number line.

1. 117

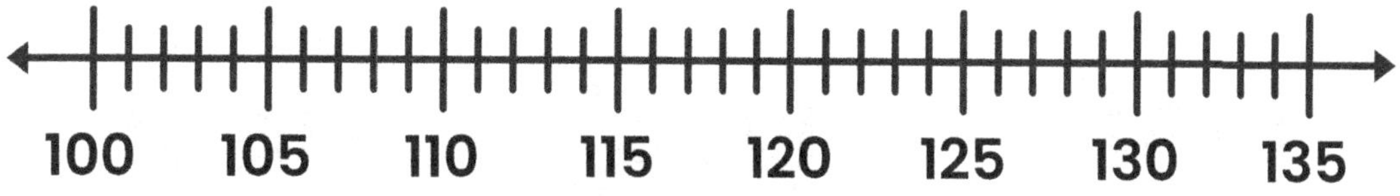


2. 649

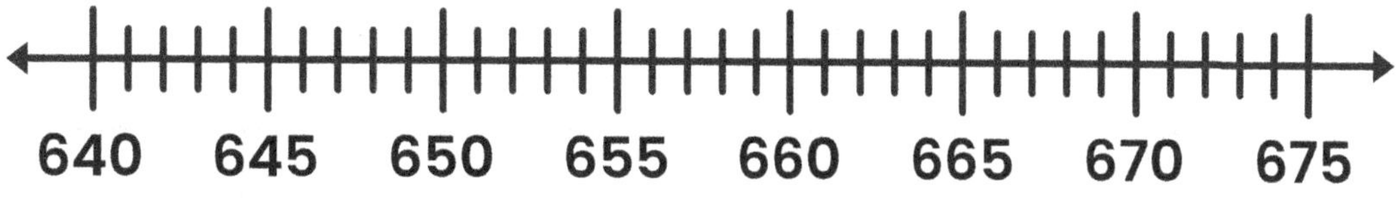


3. 843

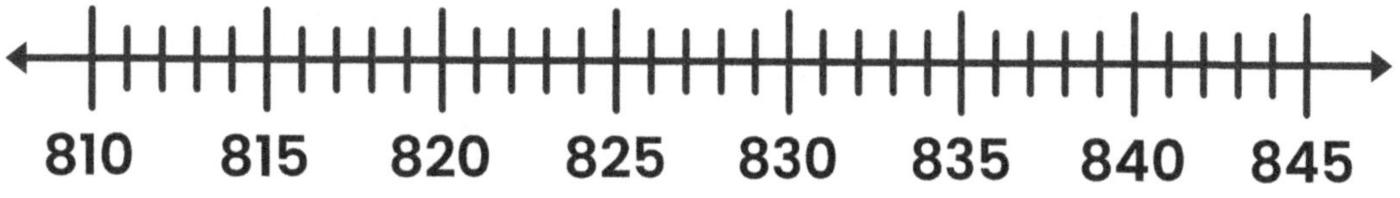


4. 958

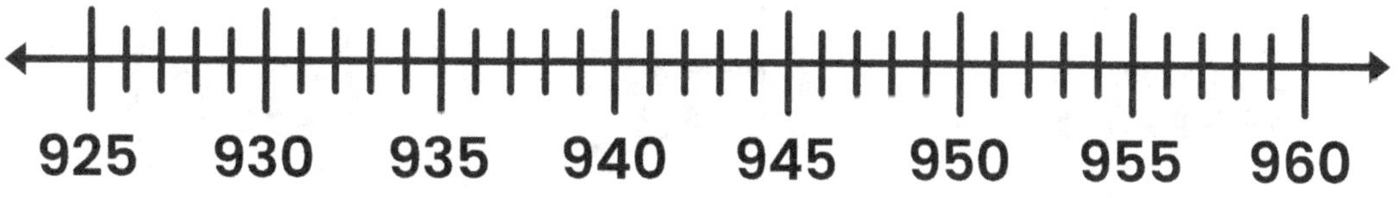


5. 502

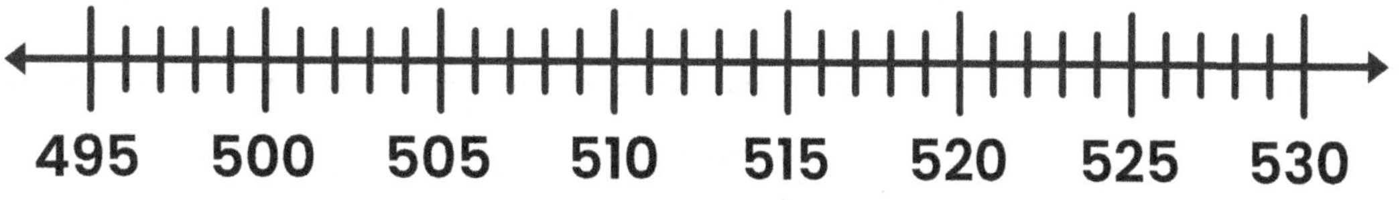


6. 299

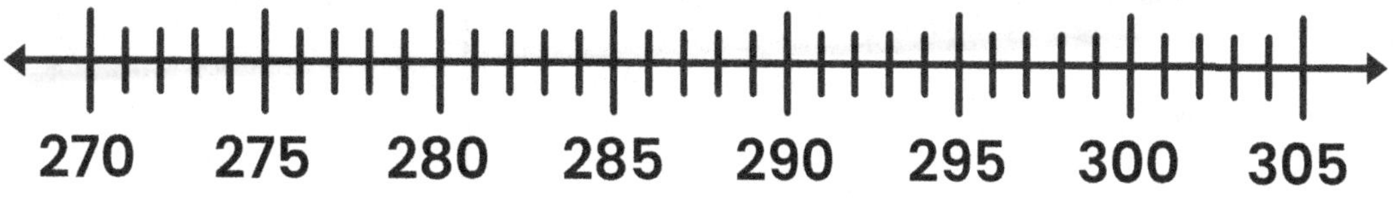

# Comparing Numbers

Draw a quick picture to represent each number. Then color the number in each pair that is **GREATER**.

| | | | |
|---|---|---|---|
| 231 | 224 | 140 | 170 |
| 381 | 390 | 540 | 550 |
| 412 | 312 | 501 | 601 |
| 240 | 214 | 340 | 430 |

# Comparing Numbers



Write the number represented by the blocks shown.Then, circle the number in each pair that is **GREATER**.

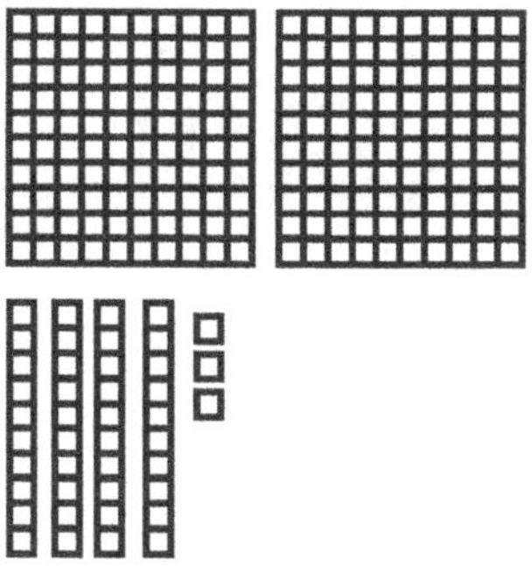

243

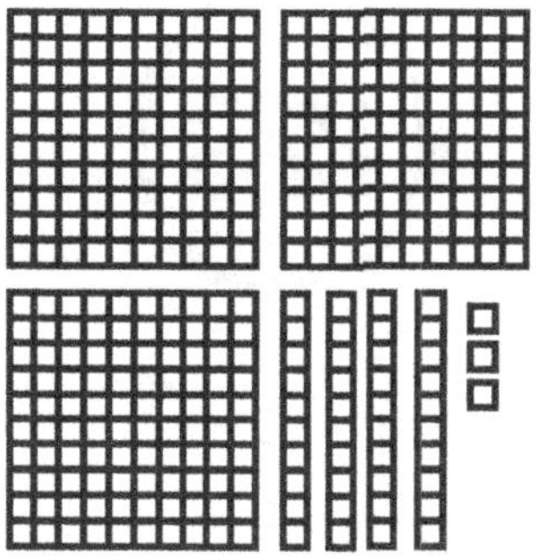

343

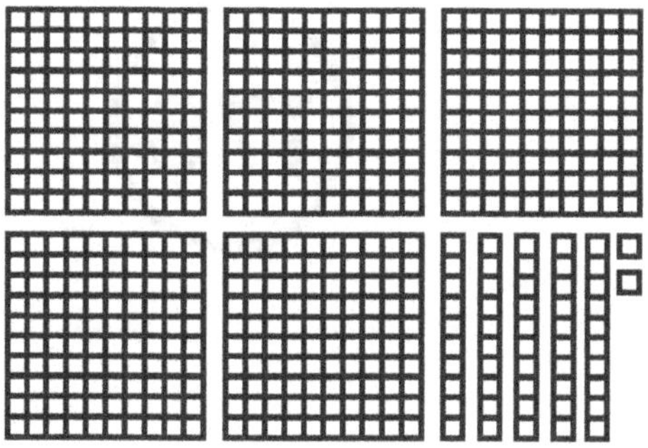

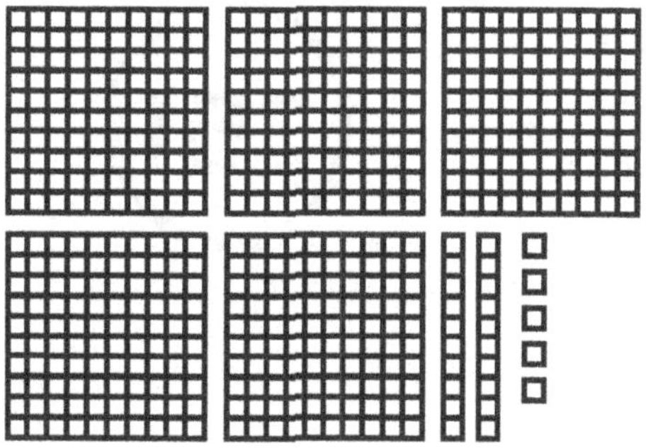

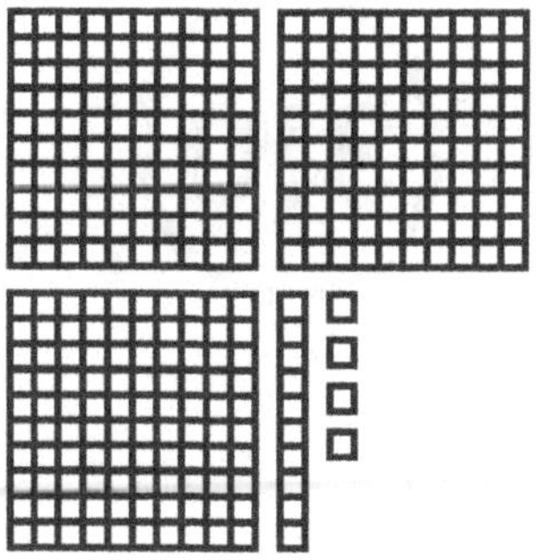

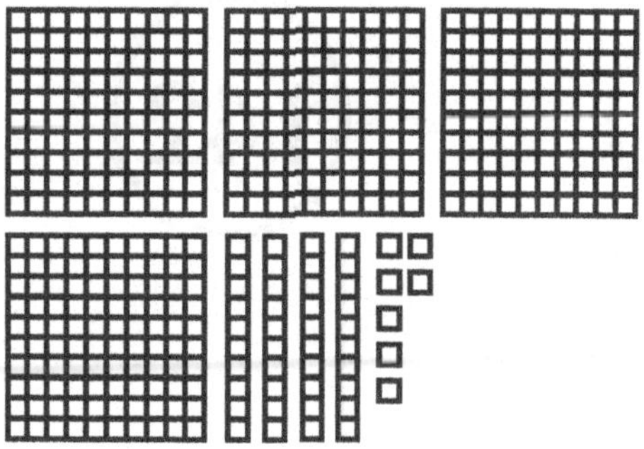

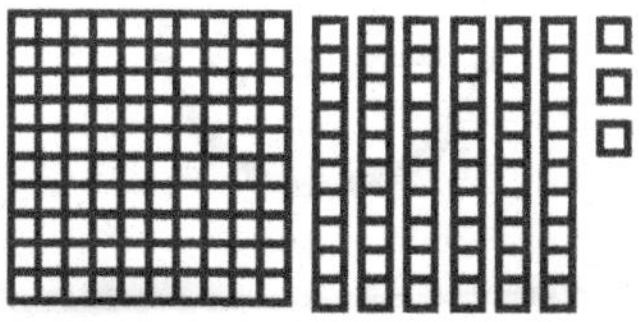

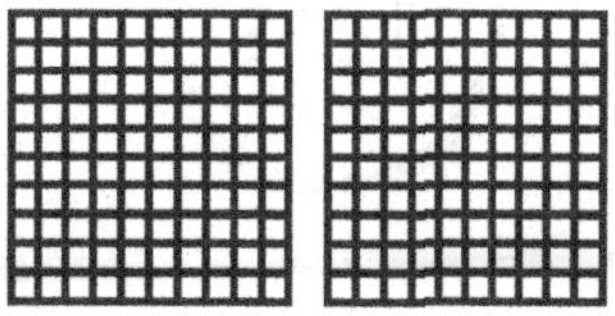

# 24 Comparing Numbers

Draw a quick picture to represent each number. Then color the number in each pair that is **LESS**.

| | |
|---|---|
| 132  123  | 410 510 |
| 341 351 | 185 175 |
| 214 142 | 301 310 |
| 117 118 | 405 413 |

Write the number represented by the blocks shown.Then, circle the number in each pair that is **LESS**.

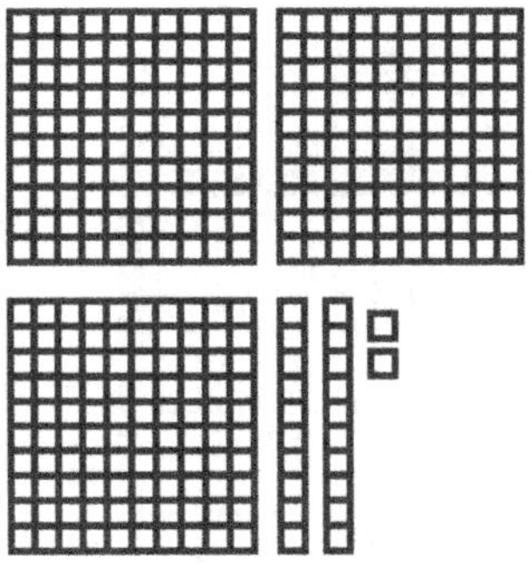

322 ______

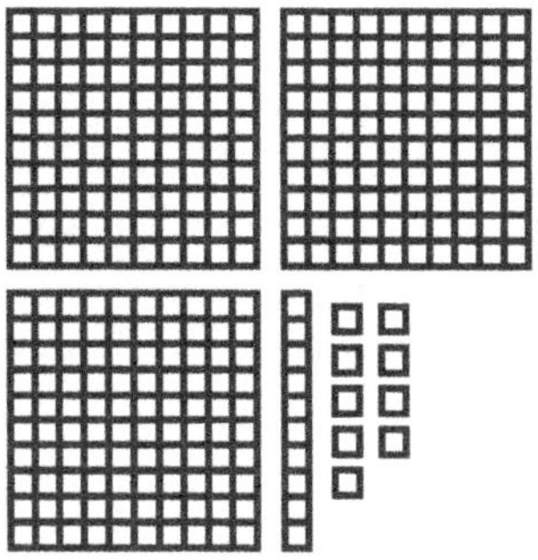

319 ______

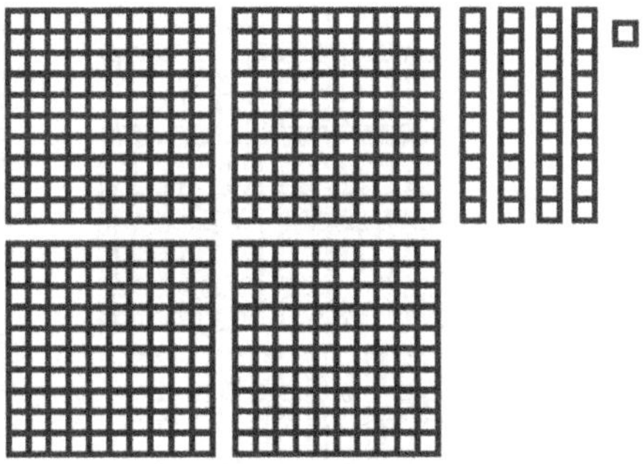

______

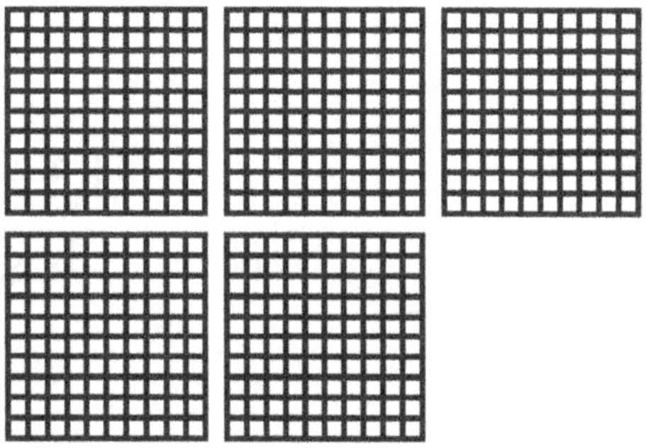

______

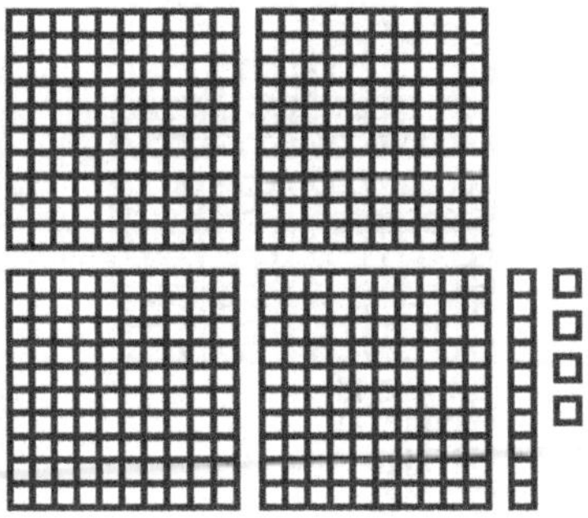

______

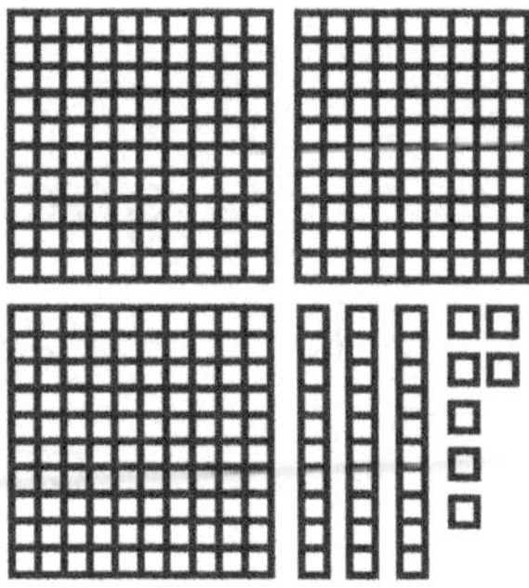

______

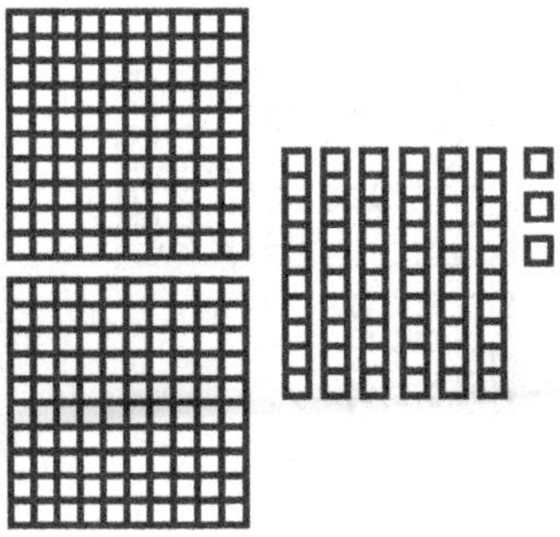

______

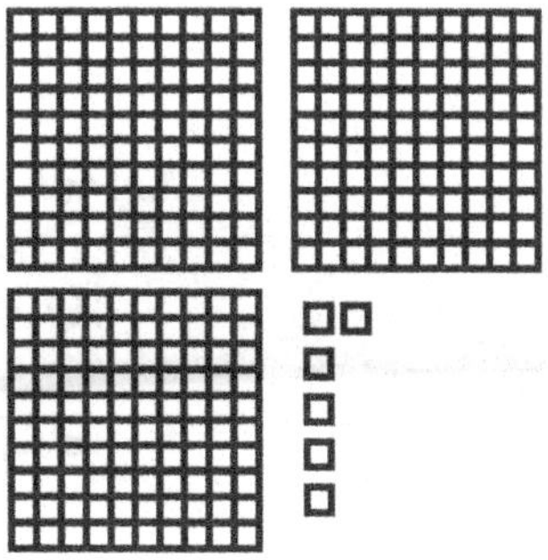

______

# Comparing Numbers

Count the blocks in each group and write the number on the line.
Then, circle the correct phrase to complete the sentence.

614 is greater than / less than / equal to 604

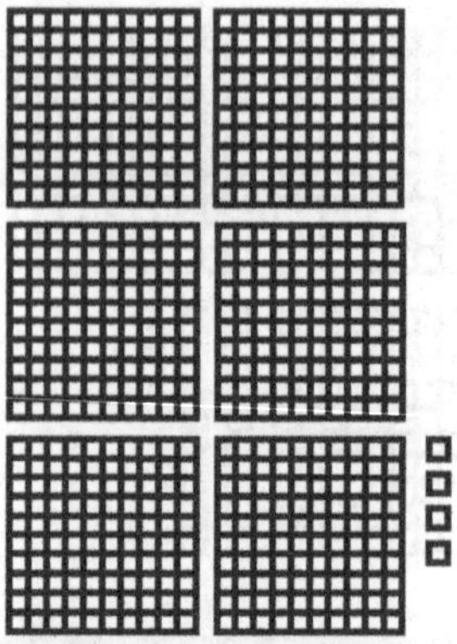

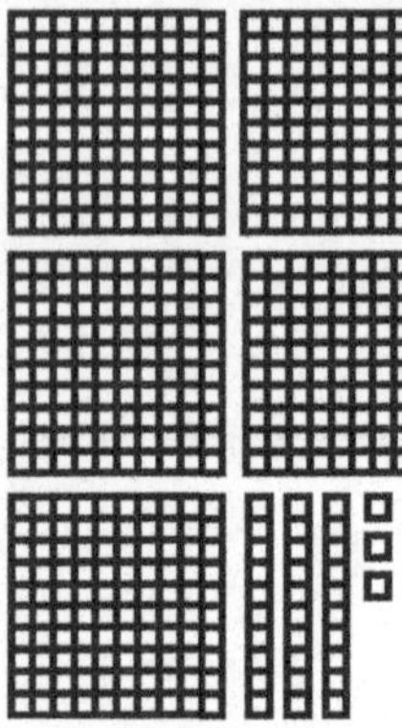

______ is greater than / less than / equal to ______

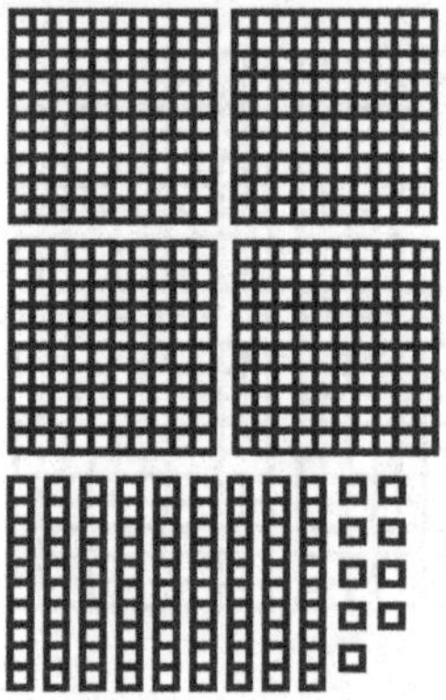

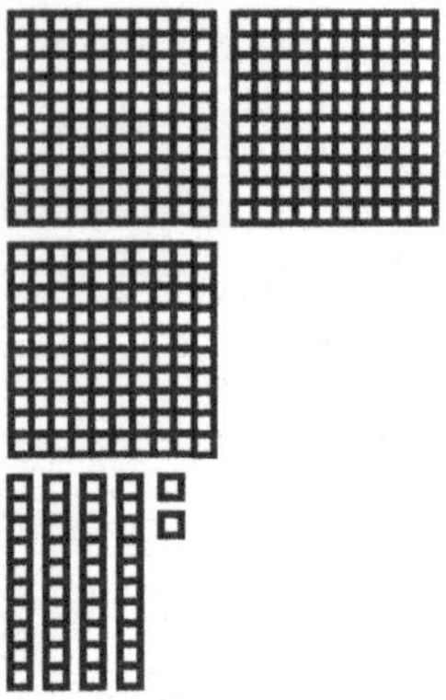

______ is greater than / less than / equal to ______

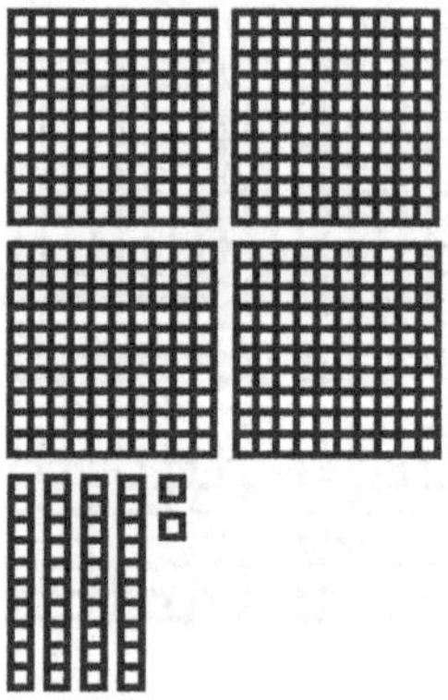

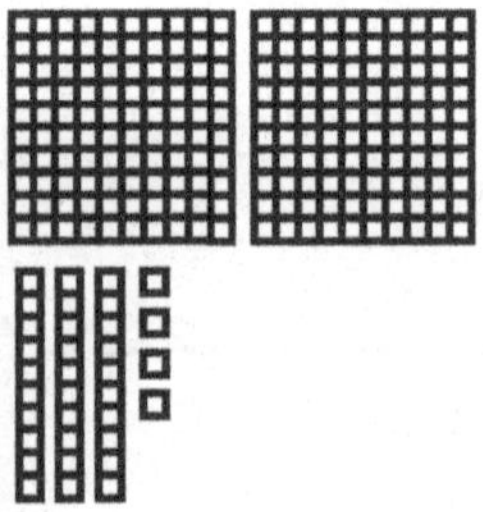

______ is greater than / less than / equal to ______

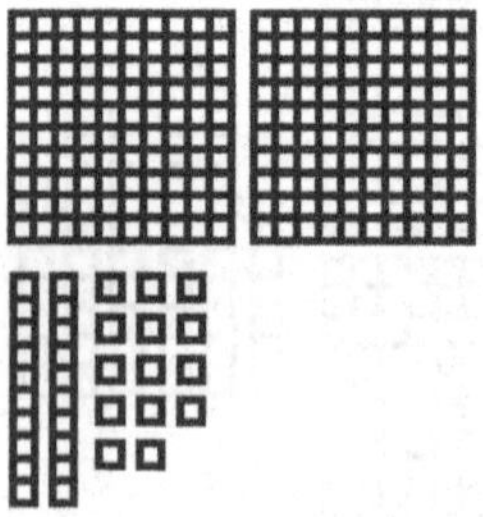

# Comparing Numbers



Compare each pair of numbers using one of these symbols:

| | | |
|---|---|---|
| > greater than | < less than | = equal to |

| | | |
|---|---|---|
| 102 ( > ) 101 | 812 (   ) 182 | 293 (   ) 294 |
| 510 (   ) 410 | 199 (   ) 199 | 351 (   ) 531 |
| 720 (   ) 720 | 524 (   ) 724 | 814 (   ) 719 |
| 218 (   ) 228 | 312 (   ) 412 | 675 (   ) 675 |
| 159 (   ) 157 | 487 (   ) 478 | 280 (   ) 525 |

# Counting Patterns

Fill in the missing numbers from 101 to 220.

| | | | | | | | | | |
|---|---|---|---|---|---|---|---|---|---|
| 101 | 102 | | 104 | | 106 | 107 | | 109 | 110 |
| 111 | | 113 | | 115 | 116 | | 118 | 119 | |
| 121 | 122 | | | 125 | 126 | 127 | | | 130 |
| | 132 | 133 | 134 | | | 137 | | 139 | 140 |
| 141 | | 143 | 144 | | 146 | | 148 | | 150 |
| 151 | 152 | | 154 | 155 | | | 158 | 159 | |
| 161 | | 163 | | | 166 | 167 | 168 | | |
| | 172 | 173 | 174 | 175 | | | 178 | | 180 |
| 181 | 182 | | 184 | | | 187 | | 189 | |
| 191 | | 193 | | | 196 | | 198 | | 200 |
| 201 | 202 | | 204 | | 206 | 207 | | 209 | |
| | | 213 | | 215 | | | 218 | | 220 |

Count on by fives and fill in the numbers that come next in each series.

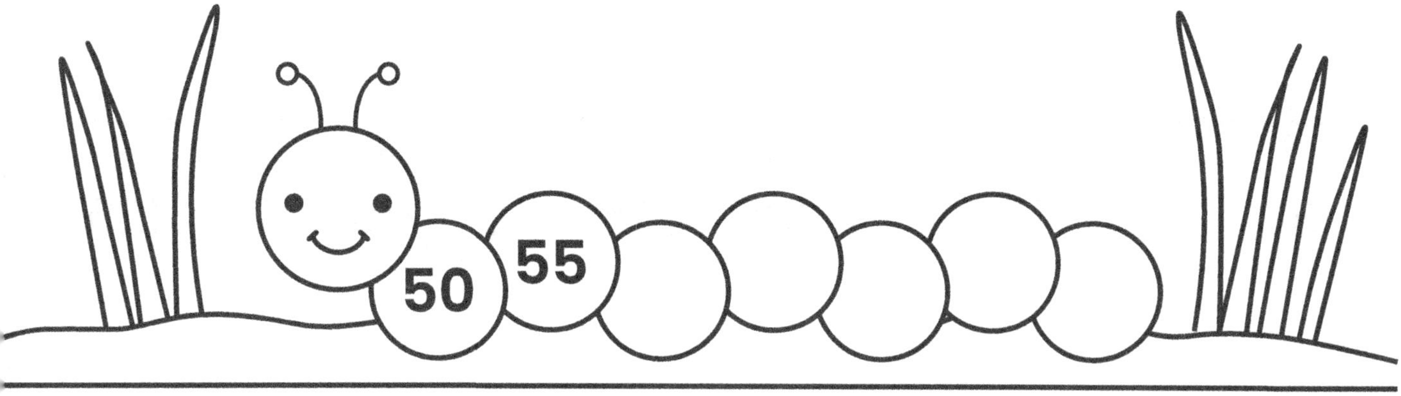


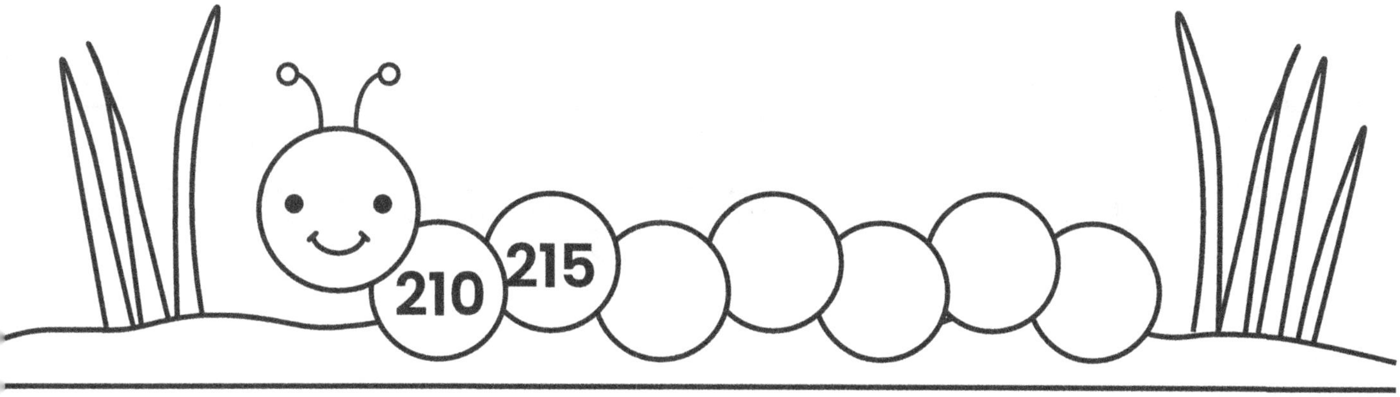


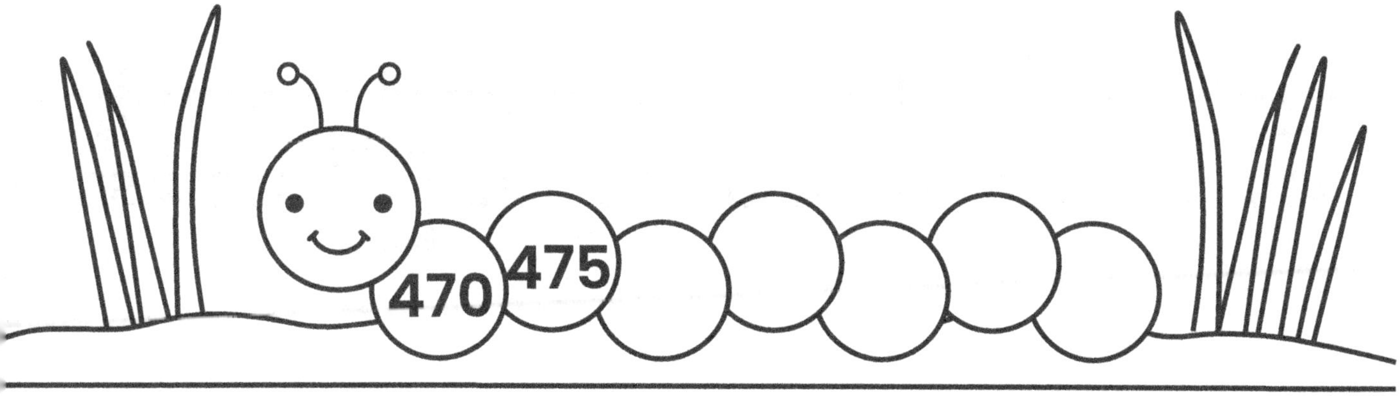


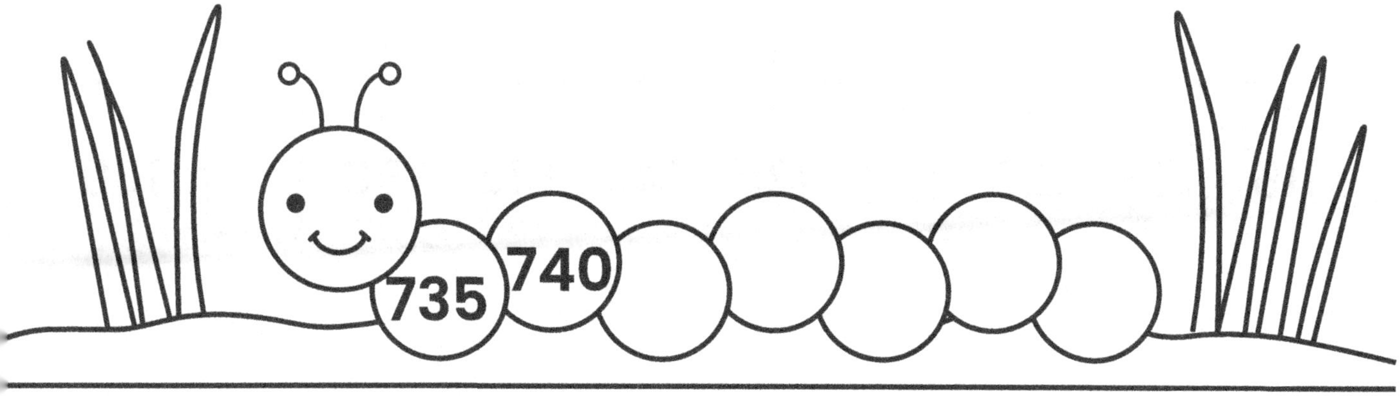

# Counting Patterns

Count by fives and fill in the missing number on each number line.

| | | | | | | | | | |
|---|---|---|---|---|---|---|---|---|---|
| 80 | 85 | 90 | | 100 | 105 | 110 | 115 | 120 | 125 |

| | | | | | | | | | |
|---|---|---|---|---|---|---|---|---|---|
| 660 | 665 | 670 | 675 | 680 | | 690 | 695 | 700 | 705 |

| | | | | | | | | | |
|---|---|---|---|---|---|---|---|---|---|
| 410 | 415 | 420 | 425 | 430 | 435 | 440 | | 450 | 455 |

| | | | | | | | | | |
|---|---|---|---|---|---|---|---|---|---|
| 115 | 120 | 125 | 130 | 135 | 140 | 145 | 150 | | 160 |

## Counting Patterns



On the hundred chart below, start at 5 and count by fives, coloring each number in the pattern as you go. Then, start at 10 and count by tens, circling each number in the pattern as you go.

| | | | | | | | | | |
|---|---|---|---|---|---|---|---|---|---|
| 1 | 2 | 3 | 4 | 5 | 6 | 7 | 8 | 9 | 10 |
| 11 | 12 | 13 | 14 | 15 | 16 | 17 | 18 | 19 | 20 |
| 21 | 22 | 23 | 24 | 25 | 26 | 27 | 28 | 29 | 30 |
| 31 | 32 | 33 | 34 | 35 | 36 | 37 | 38 | 39 | 40 |
| 41 | 42 | 43 | 44 | 45 | 46 | 47 | 48 | 49 | 50 |
| 51 | 52 | 53 | 54 | 55 | 56 | 57 | 58 | 59 | 60 |
| 61 | 62 | 63 | 64 | 65 | 66 | 67 | 68 | 69 | 70 |
| 71 | 72 | 73 | 74 | 75 | 76 | 77 | 78 | 79 | 80 |
| 81 | 82 | 83 | 84 | 85 | 86 | 87 | 88 | 89 | 90 |
| 91 | 92 | 93 | 94 | 95 | 96 | 97 | 98 | 99 | 100 |

# Counting Patterns

For each train, count on by tens from the number on the engine and write a number in each train car.

# Counting Patterns



Write the numbers that are ten less and ten more than the number shown.

| 10 Less | | 10 More |
|---|---|---|
| 210 | **220** | 230 |

| 10 Less | | 10 More |
|---|---|---|
| | **350** | |

| 10 Less | | 10 More |
|---|---|---|
| | **830** | |

| 10 Less | | 10 More |
|---|---|---|
| | **160** | |

| 10 Less | | 10 More |
|---|---|---|
| | **600** | |

| 10 Less | | 10 More |
|---|---|---|
| | **910** | |

| 10 Less | | 10 More |
|---|---|---|
| | **780** | |

| 10 Less | | 10 More |
|---|---|---|
| | **470** | |

| 10 Less | | 10 More |
|---|---|---|
| | **590** | |

| 10 Less | | 10 More |
|---|---|---|
| | **140** | |

# 34 Counting Patterns

Count by hundreds. Write the numbers in the clouds.

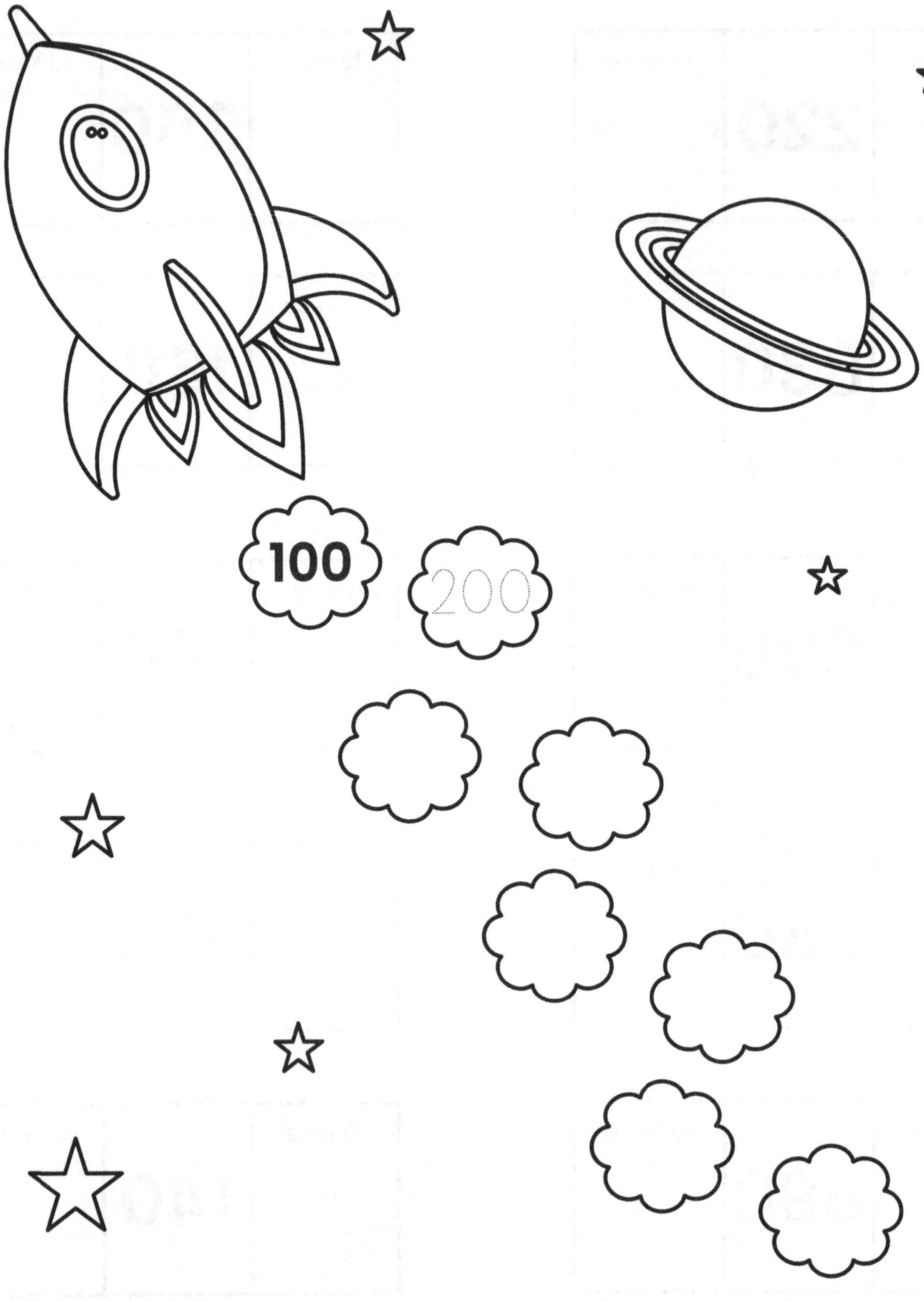

Count by hundreds and fill in the missing number on each number line.

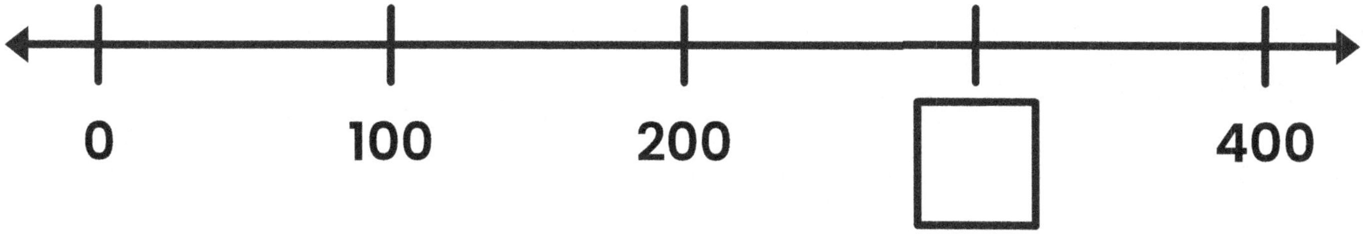


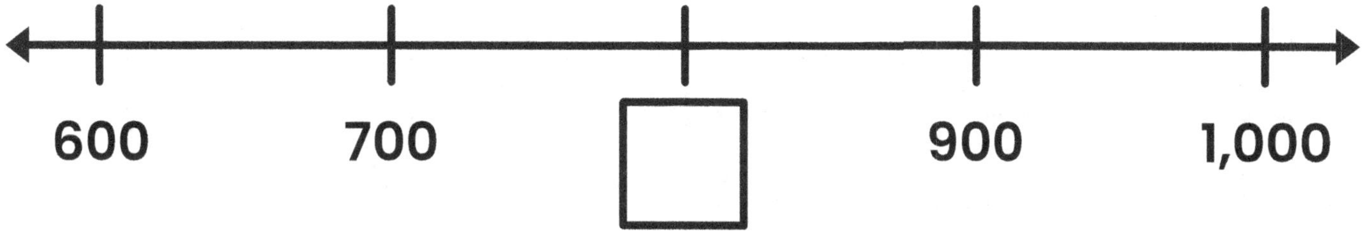


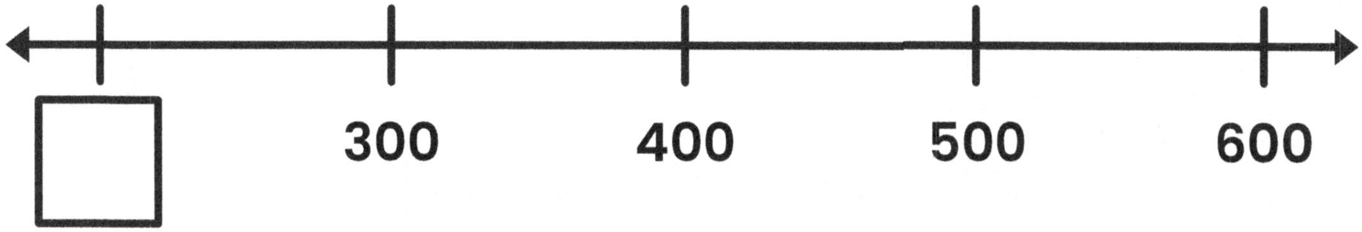


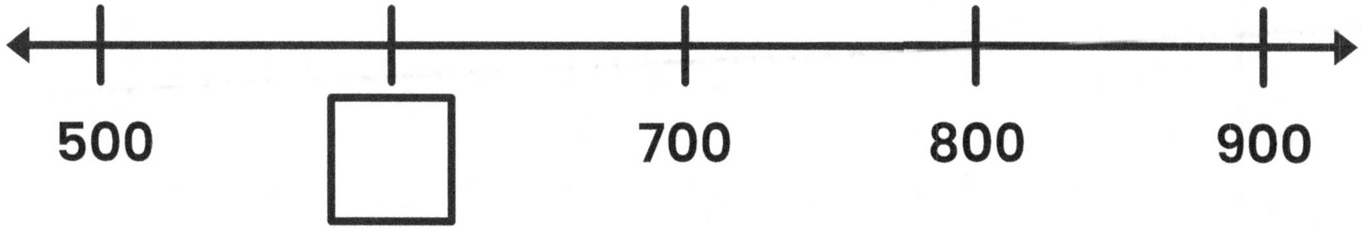


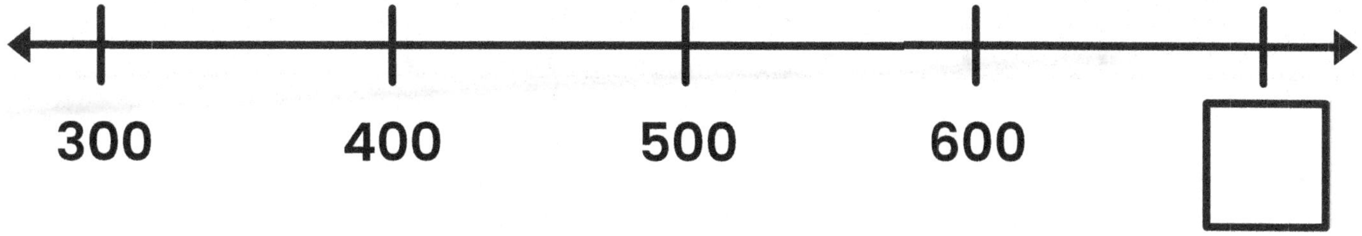

# Counting Patterns

Fill in the missing numbers on the hundred chart below.

| | | | | | | | | | |
|---|---|---|---|---|---|---|---|---|---|
| 501 | 502 | 503 | 504 | 505 | 506 | 507 | 508 | 509 | |
| 511 | 512 | 513 | 514 | 515 | 516 | 517 | 518 | 519 | |
| 521 | 522 | 523 | 524 | 525 | 526 | 527 | 528 | 529 | |
| 531 | 532 | 533 | 534 | 535 | 536 | 537 | 538 | 539 | |
| 541 | 542 | 543 | 544 | 545 | 546 | 547 | 548 | 549 | |
| 551 | 552 | 553 | 554 | 555 | 556 | 557 | 558 | 559 | |
| 561 | 562 | 563 | 564 | 565 | 566 | 567 | 568 | 569 | |
| 571 | 572 | 573 | 574 | 575 | 576 | 577 | 578 | 579 | |
| 581 | 582 | 583 | 584 | 585 | 586 | 587 | 588 | 589 | |
| 591 | 592 | 593 | 594 | 595 | 596 | 597 | 598 | 599 | |

The numbers along each kite tail follow a counting pattern. Fill in the missing numbers on each kite tail.

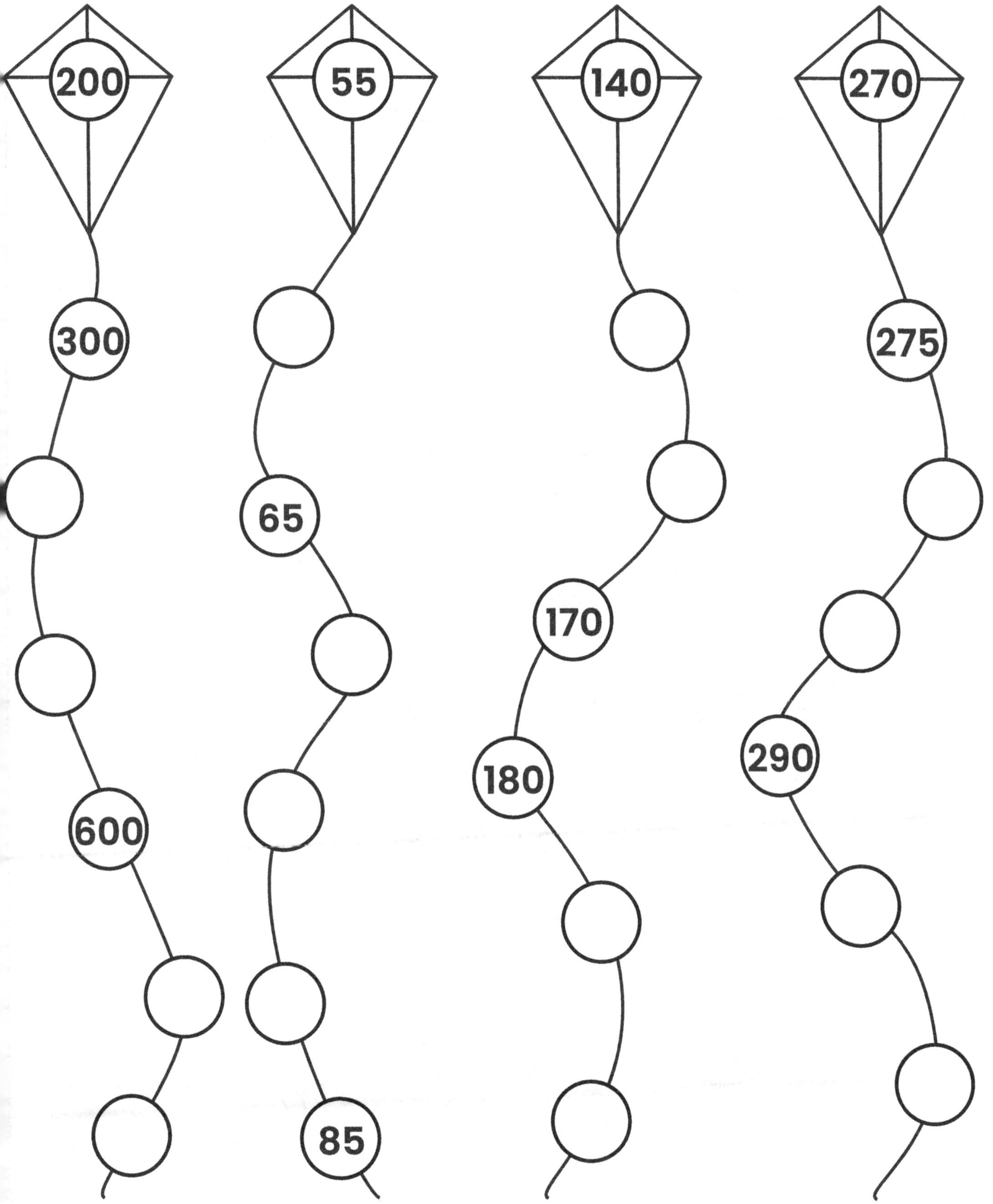

# Evens and Odds

Circle pairs of bugs. Write the number and circle whether it is even or odd.

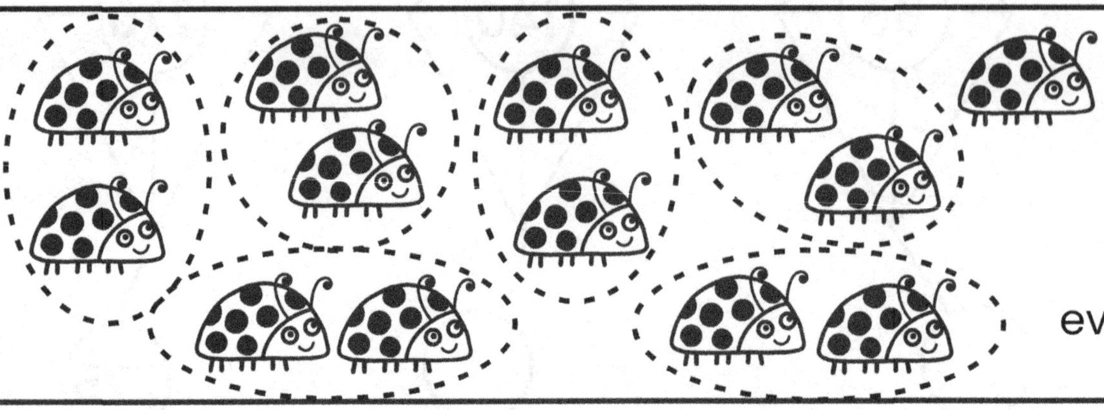

13

even odd

even odd

even odd

even odd

Draw to represent each number. Circle whether the number is even or odd.

| | |
|---|---|
| 15 even odd | 9 even odd |
| 18 even odd | 12 even odd |
| 11 even odd | 14 even odd |
| 16 even odd | 7 even odd |

# 40 Evens and Odds

Color the apples with odd numbers **RED** and the apples with even numbers **GREEN**.

Fill in the correct bubble to answer each question.

Which of these shows an odd number of blocks?

○ 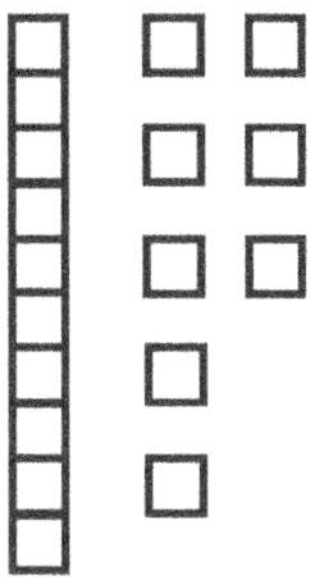   ○    ○ 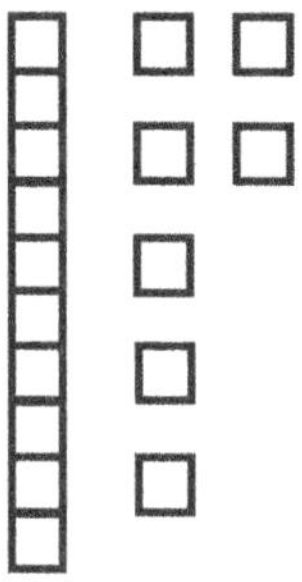

Which of these shows an even number of blocks?

○ 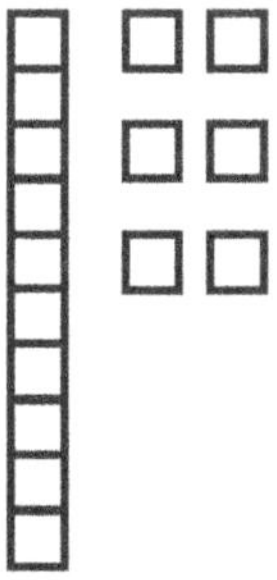   ○ 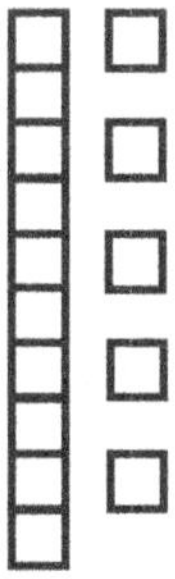   ○ 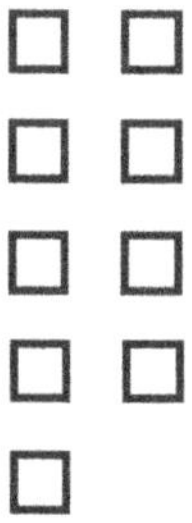

Which of these is an even number less than 10?

○ 3   ○ 12   ○ 8

Which of these is an odd number greater than 10?

○ 15   ○ 14   ○ 9

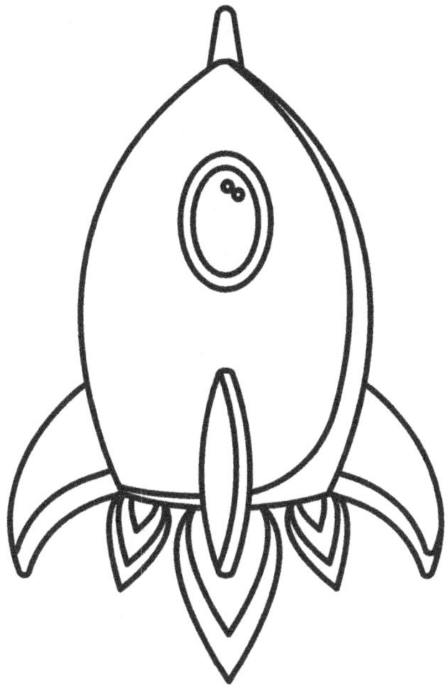

# Operations

# Add and Subtract Within 20

Use the ten frames to show each addition problem. Then, write the solution.

7 + 4 = 11

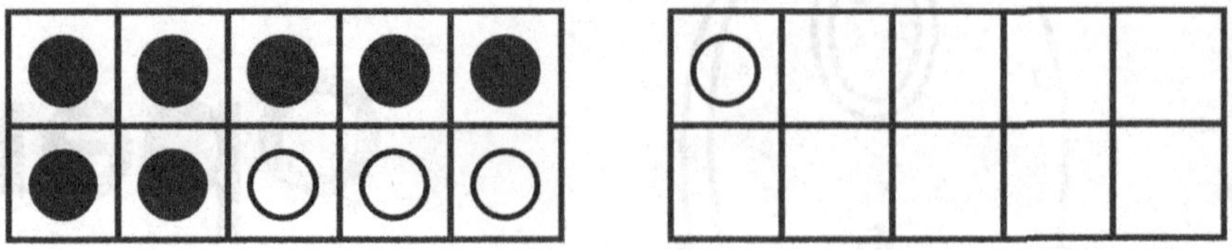

8 + 2 = ____

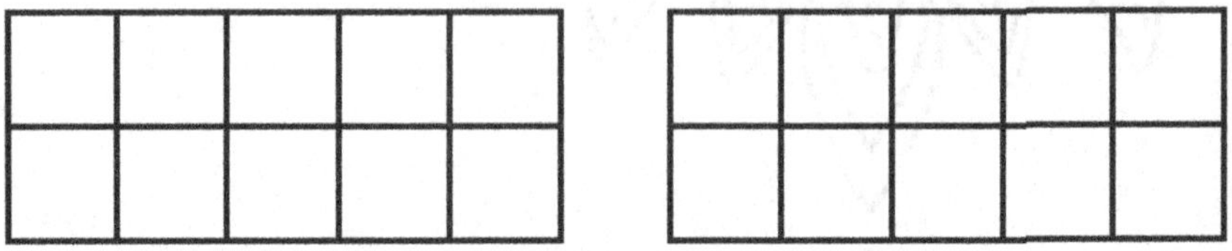

3 + 9 = ____

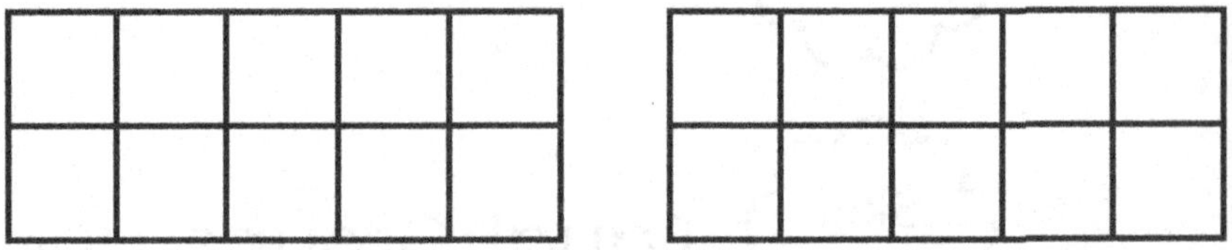

10 + 6 = ____

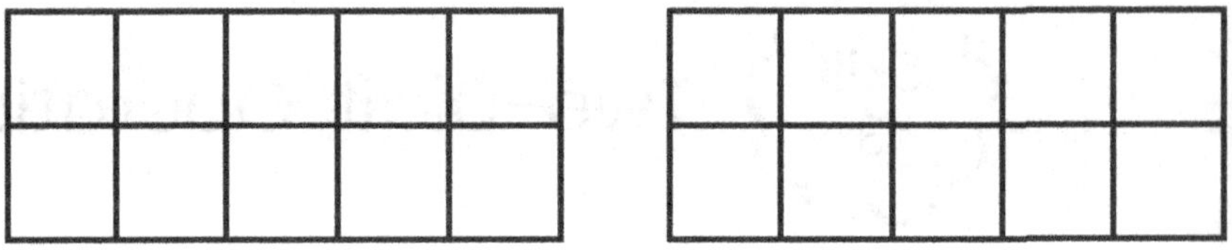

3 + 4 = ____

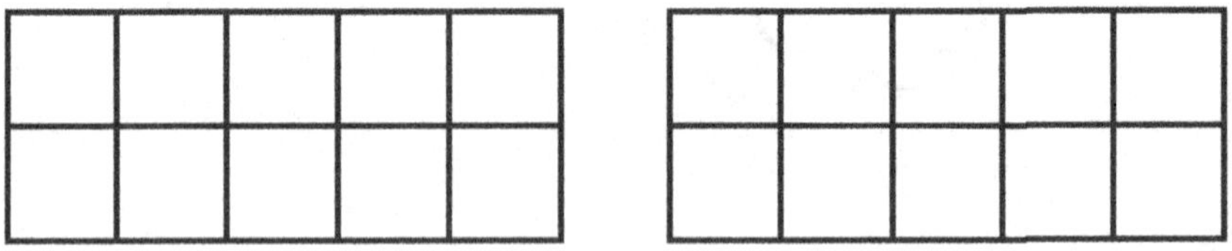

6 + 7 = ____

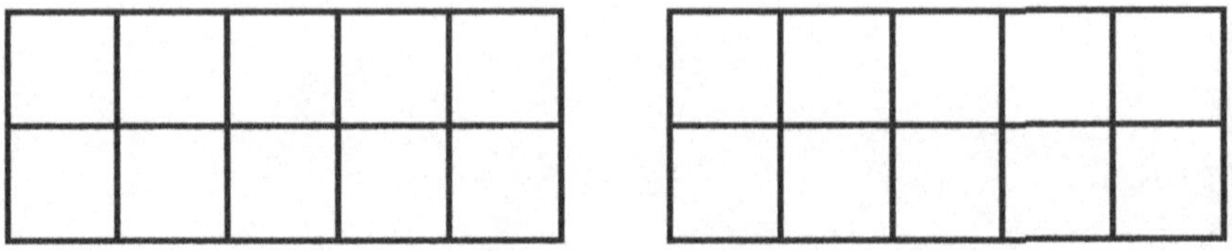

# Add and Subtract Within 20



Choose the equation in each group that is best represented by the ten frames shown.

1

- ○ 7 + 5 = 13
- ○ 8 + 5 = 13
- ○ 8 + 2 = 10

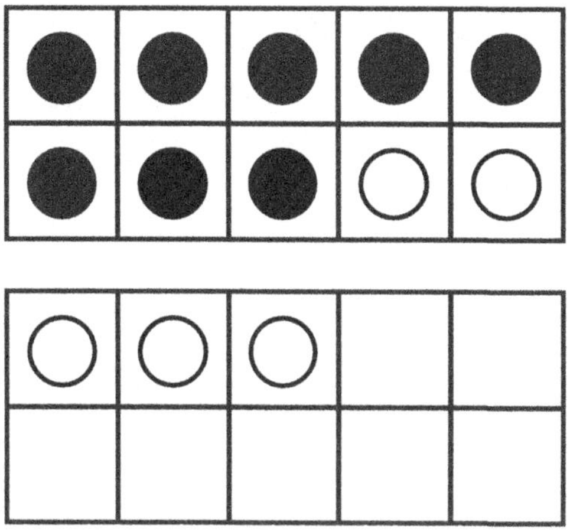

2

- ○ 11 = 5 + 6
- ○ 5 + 5 = 10
- ○ 5 + 6 = 12

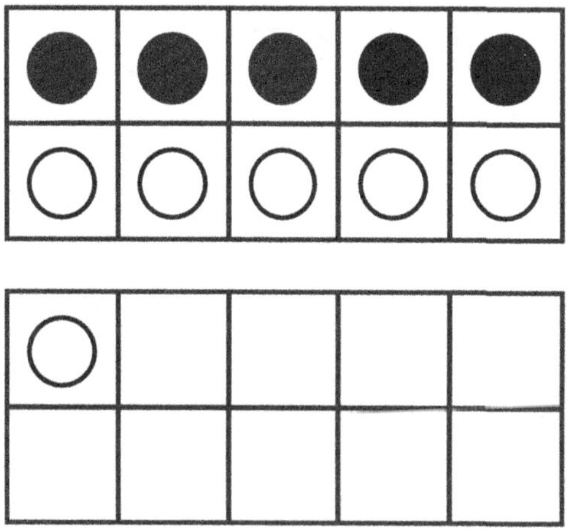

3

- ○ 12 = 10 + 2
- ○ 6 + 4 = 10
- ○ 6 + 6 = 12

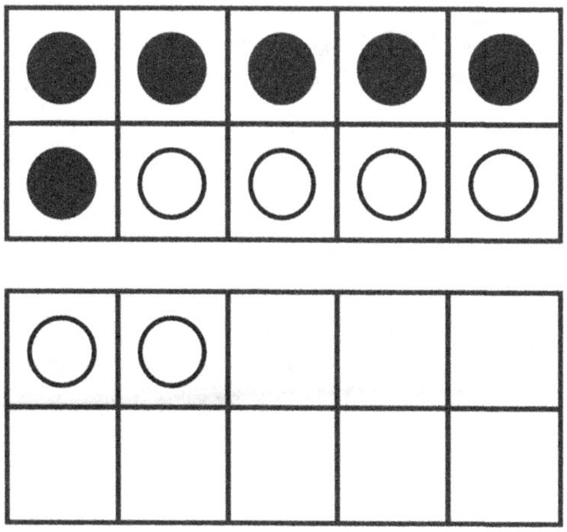

Draw a picture to help you solve each addition equation.

| | |
|---|---|
| 8 + 4 = ____ | 12 + 3 = ____ |
| 9 + 1 = ____ | 6 + 2 = ____ |
| 5 + 6 = ____ | 10 + 3 = ____ |

Use the number line to solve each addition equation.

7 + 2 = 9

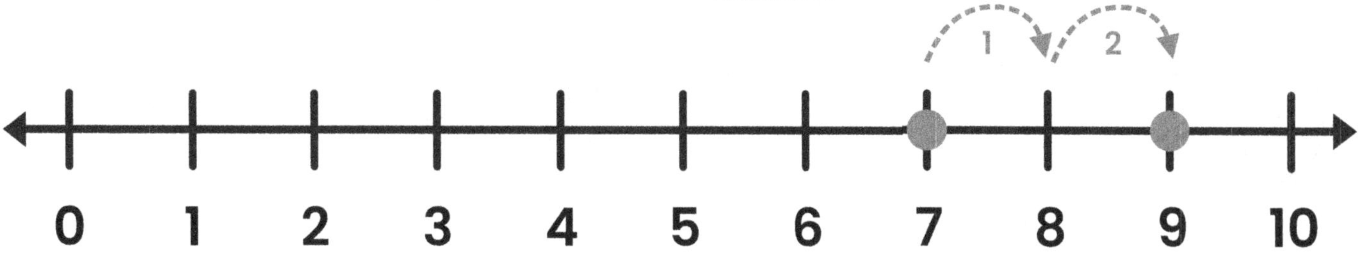


4 + 5 = ____

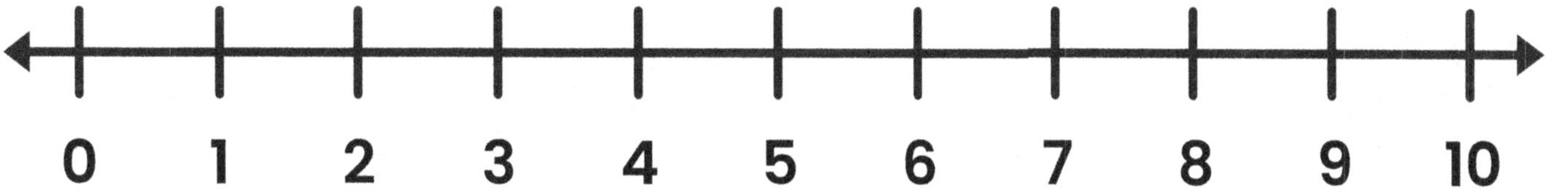


11+ 3 = ____

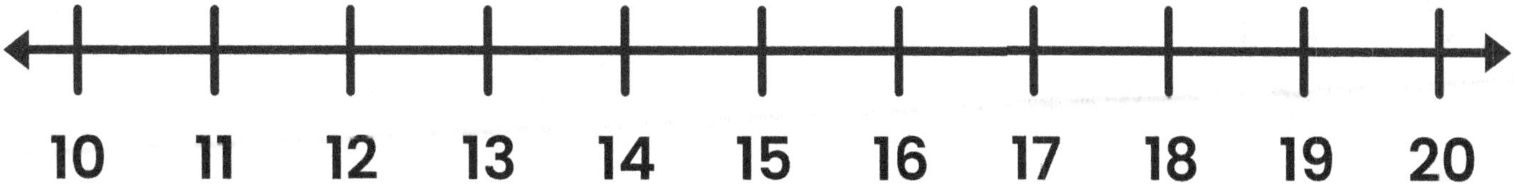


14 + 4 = ____

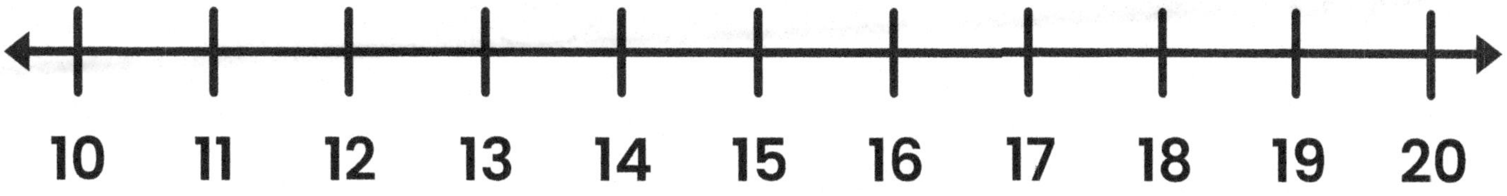

Use doubles facts to add the numbers in each pair of cherries.

____ 2

____

____

____

____

____

____

____

# Add and Subtract Within 20



Use doubles facts to help you solve each addition problem.

**EX**

8 + 9 = ?

8 + 8 + 1 = ?

16 + 1 = 17

8 + 9 = 17

**1**

5 + 6 = ?

____ + ____ + ____ = ?

____ + ____ = ____

5 + 6 = ____

**2**

7 + 8 = ?

____ + ____ + ____ = ?

____ + ____ = ____

7 + 8 = ____

**3**

6 + 8 = ?

____ + ____ + ____ = ?

____ + ____ = ____

6 + 8 = ____

# Add and Subtract Within 20

Use circles to represent the addition problem in and below the ten frame. Then, use the strategy of making a ten to solve.

EX

**7 + 5 = ?**

10 + 2 = 12

7 + 5 = 12

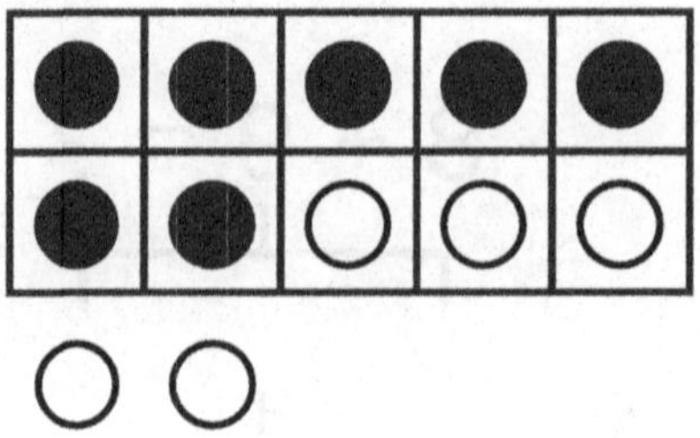

1

**4 + 8 = ?**

10 + ____ = ____

4 + 8 = ____

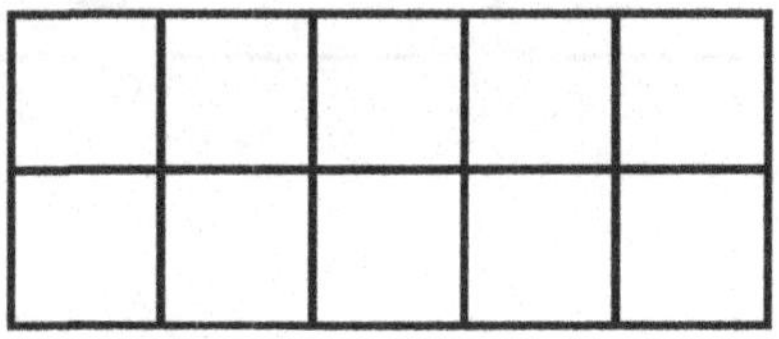

2

**9 + 7 = ?**

10 + ____ = ____

9 + 7 = ____

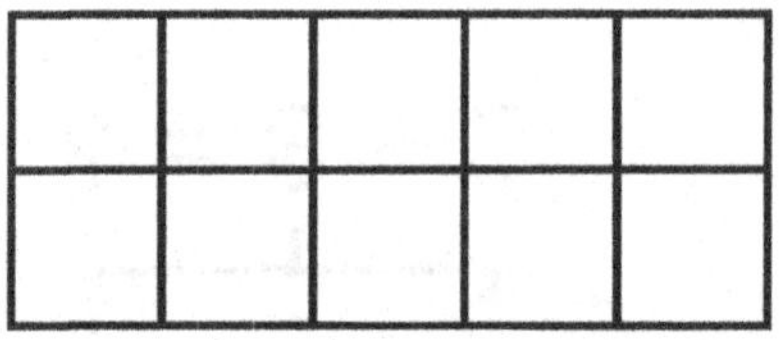

3

**6 + 7 = ?**

10 + ____ = ____

6 + 7 = ____

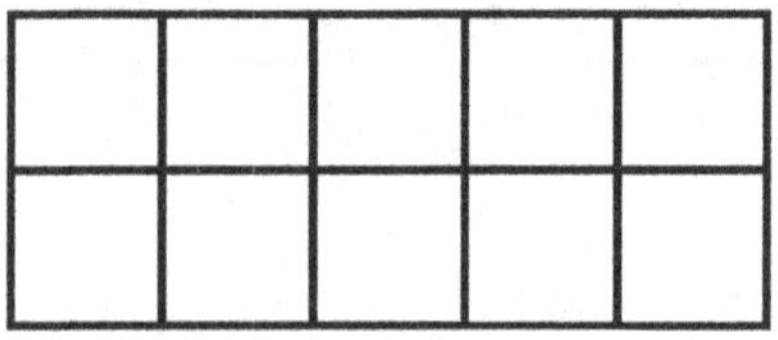

Use the strategy of making a ten to solve each addition problem.

**EX**

7 + 5 = ?

7 + 3 + 2 = ?

10 + 2 = 12

7 + 5 = 12

**1**

8 + 6 = ?

____ + ____ + ____ = ?

____ + ____ = ____

8 + 6 = ____

**2**

9 + 8 = ?

____ + ____ + ____ = ?

____ + ____ = ____

9 + 8 = ____

**3**

6 + 9 = ?

____ + ____ + ____ = ?

____ + ____ = ____

6 + 9 = ____

# Add and Subtract Within 20

Read each word problem. Draw a picture in the box, then fill in the numbers to complete the equation.

Nina has 2 sisters and 3 brothers. How many siblings does Nina have?

_____ + _____ = _____

Luke scored 8 points in the first half of the basketball game and 4 points in the second half. How many points did he score in all?

_____ + _____ = _____

Nadia's backyard has 3 oak trees and 4 maple trees. How many trees are in her backyard?

_____ + _____ = _____

Read each word problem. Draw a picture in the box, then fill in the numbers to complete the equation.

| | |
|---|---|
| Carlos rolls a pair of dice during a board game. One of the dice shows a 6 and the other shows a 4. What is the total shown on both dice?<br><br>_____ + _____ = _____ | |
| Amanda is 11. Her brother is 2 years older than her. How old is Amanda's brother?<br><br>_____ + _____ = _____ | |
| Jason read 15 pages of a book during silent reading time. Kelly read 3 more pages than Jason did. How many pages did Kelly read?<br><br>_____ + _____ = _____ | |

Find each sum.

| | | | | | |
|---|---|---|---|---|---|
| 5<br>+ 1<br>6 | 2<br>+ 2 | 3<br>+ 3 | 12<br>+ 3 | 8<br>+ 4 | 5<br>+ 7 |
| 4<br>+ 2 | 11<br>+ 2 | 5<br>+ 3 | 14<br>+ 3 | 10<br>+ 4 | 9<br>+ 5 |
| 1<br>+ 9 | 10<br>+ 9 | 3<br>+ 7 | 16<br>+ 3 | 9<br>+ 9 | 11<br>+ 5 |
| 8<br>+ 2 | 15<br>+ 2 | 9<br>+ 3 | 4<br>+ 4 | 14<br>+ 4 | 13<br>+ 5 |
| 10<br>+ 2 | 17<br>+ 2 | 8<br>+ 8 | 6<br>+ 4 | 5<br>+ 6 | 6<br>+ 6 |

Find each sum.

| | | | | | |
|---|---|---|---|---|---|
| 12 | 2 | 3 | 4 | 5 | 7 |
| + 2 | + 3 | + 8 | + 7 | + 5 | + 6 |
| | | | | | |
| 2 | 14 | 11 | 9 | 5 | 8 |
| + 5 | + 2 | + 3 | + 4 | + 8 | + 6 |
| | | | | | |
| 7 | 16 | 13 | 11 | 10 | 7 |
| + 2 | + 2 | + 3 | + 4 | + 5 | + 8 |
| | | | | | |
| 2 | 3 | 15 | 13 | 12 | 9 |
| + 9 | + 4 | + 3 | + 4 | + 5 | + 8 |
| | | | | | |
| 2 | 6 | 4 | 15 | 14 | 10 |
| + 4 | + 3 | + 5 | + 4 | + 5 | + 1 |

# Add and Subtract Within 20

Use the ten frames to show each subtraction problem. Then, write the solution.

12 - 1 = 11

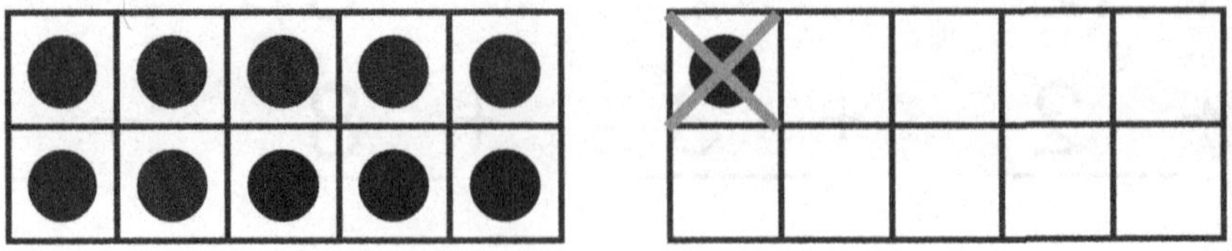

17 - 2 = ____

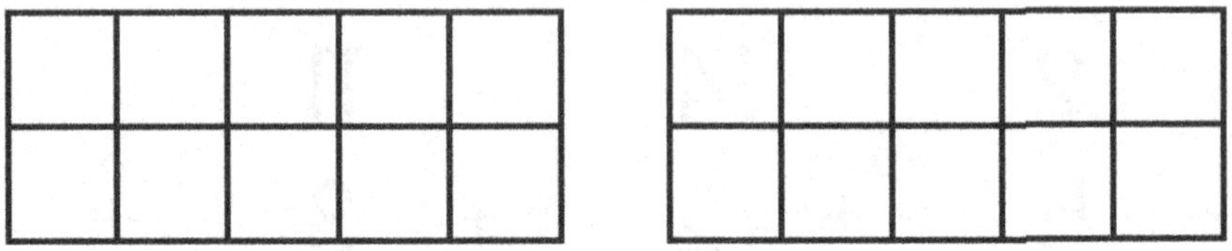

10 - 1 = ____

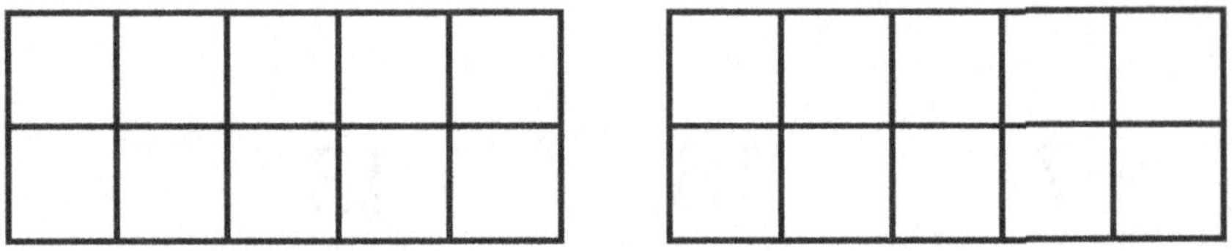

15 - 4 = ____

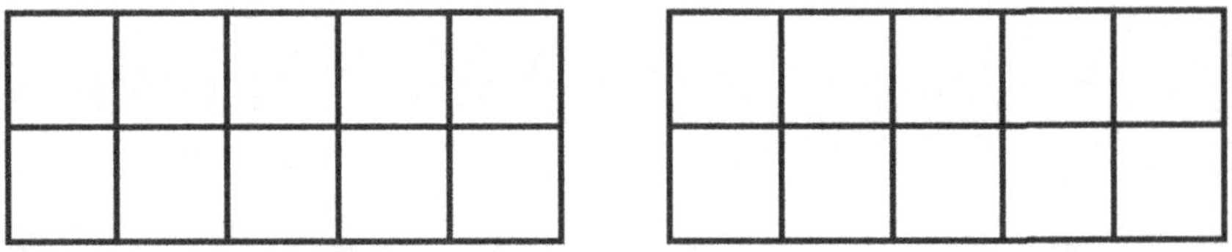

11 - 9 = ____

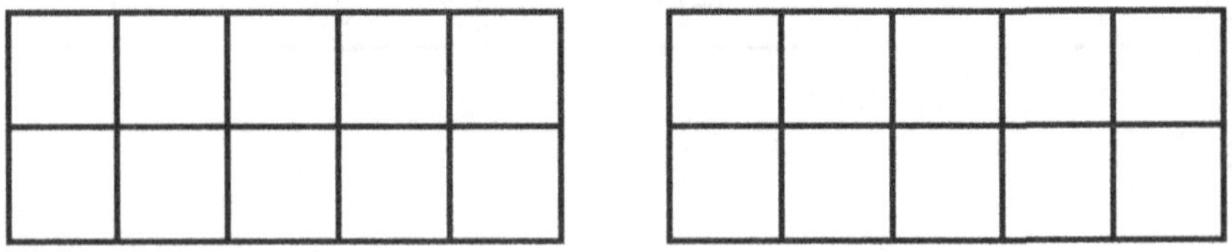

6 - 4 = ____

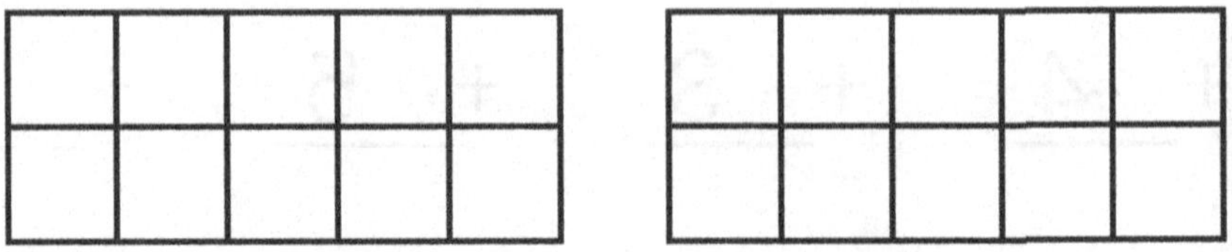

Choose the subtraction equation in each group that is best represented by the ten frames shown.

1.

- ○ 13 - 3 = 10
- ○ 10 - 1 = 9
- ○ 13 - 4 = 9

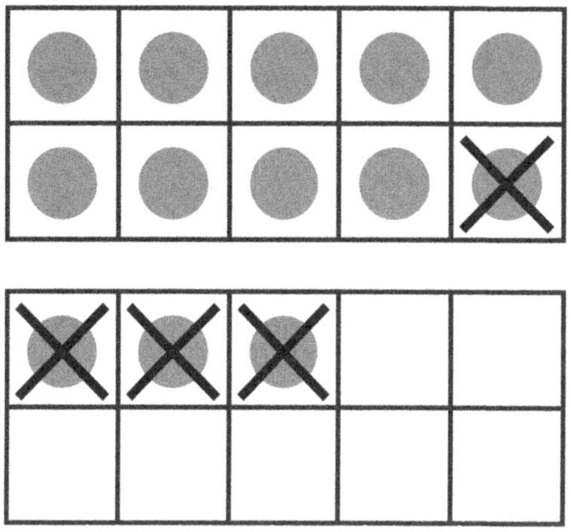

2.

- ○ 4 = 14 - 4
- ○ 14 = 10 - 4
- ○ 10 = 14 - 4

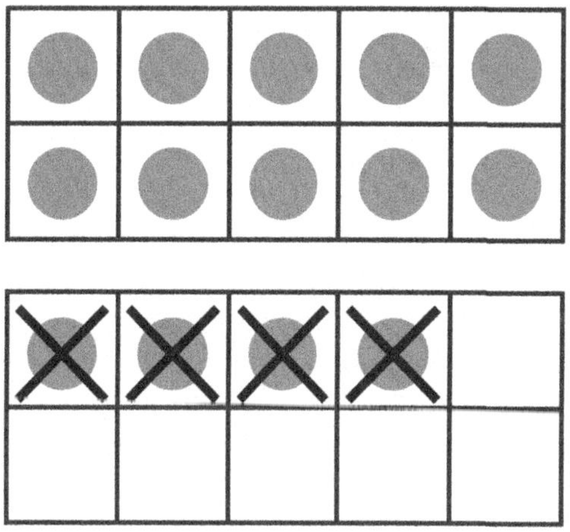

3.

- ○ 12 = 7 - 4
- ○ 10 - 3 = 7
- ○ 12 - 5 = 7

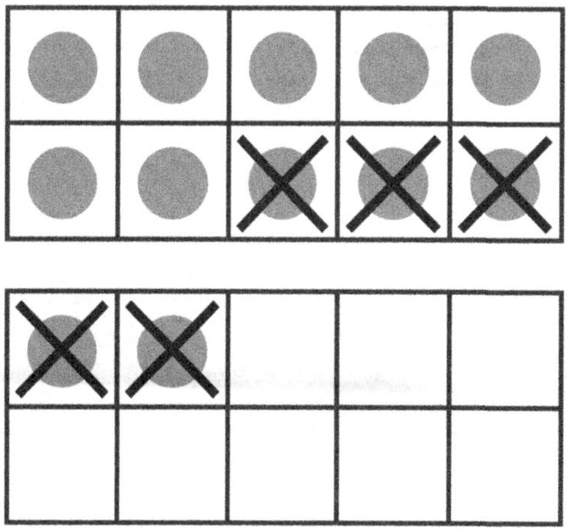

# Add and Subtract Within 20

Draw a picture to help you solve each subtraction equation.

| | |
|---|---|
| 8 - 4 = ____ | 10 - 3 = ____ |
| 19 - 1 = ____ | 16 - 2 = ____ |
| 15 - 6 = ____ | 11 - 4 = ____ |

# Add and Subtract Within 20



Use the number line to solve each subtraction equation.

7 - 2 = 5

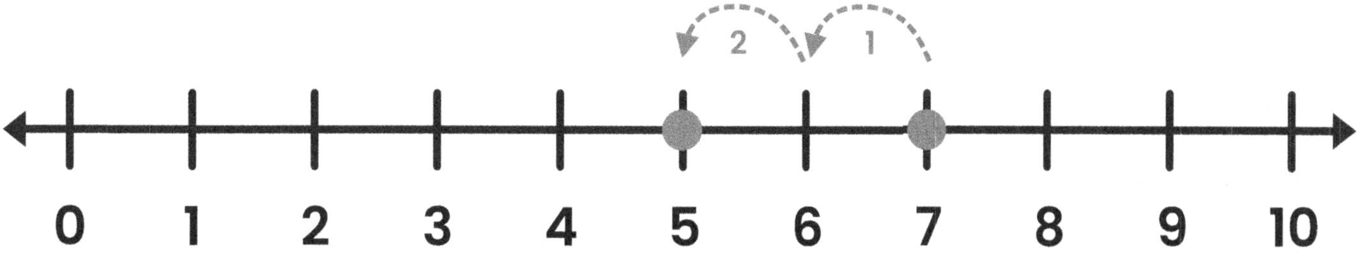


9 - 5 = ____

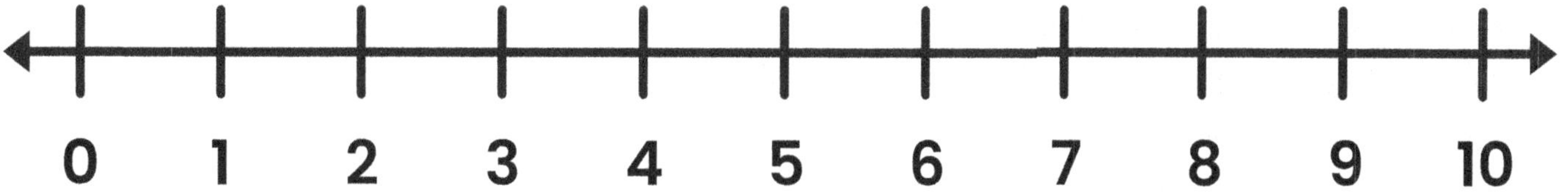


20 - 3 =____

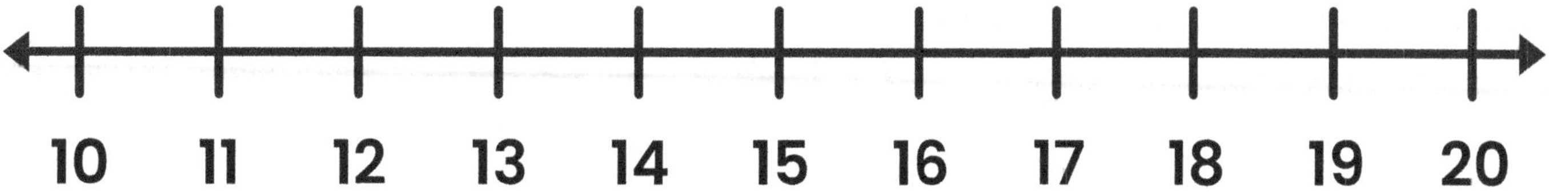


16 - 4 =____

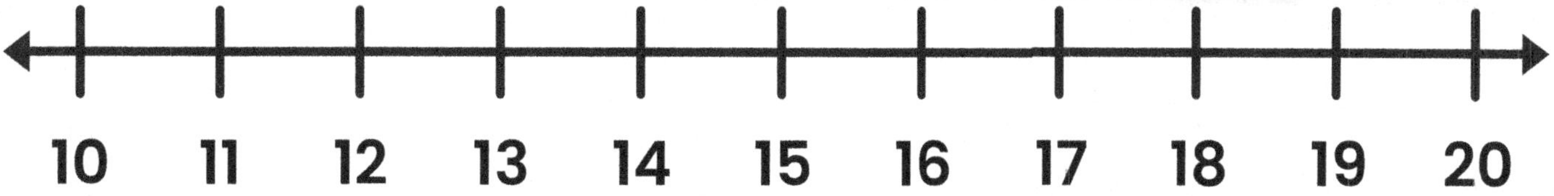

# Add and Subtract Within 20

Read each word problem. Draw a picture in the box, then fill in the numbers to complete the equation.

| | |
|---|---|
| There are 15 questions on a quiz. Mary has already answered 4 of them. How many questions does she have left to answer?<br><br>_____ - _____ = _____ | |
| Priya brought a pack of 18 cupcakes to share with her class. After each student takes one, there are 2 cupcakes left over. How many students are in the class?<br><br>_____ - _____ = _____ | |
| Ethan is 10 years old. He is 3 years older than his sister Claire. How old is Claire?<br><br>_____ - _____ = _____ | |

Read each word problem. Draw a picture in the box, then fill in the numbers to complete the equation.

| | |
|---|---|
| Luis has 15 dollars. He spends 5 dollars on a toy. How much money does he have left?<br><br>_____ - _____ = _____ | |
| There are 12 chapters in a book. Derek has read 4 chapters so far. How many chapters are left?<br><br>_____ - _____ = _____ | |
| Mae checked 8 books out of the library. 4 of the books are fiction. How many of the books are non-fiction?<br><br>_____ - _____ = _____ | |

# Add and Subtract Within 20

Find each difference.

| | | | | | |
|---|---|---|---|---|---|
| 5 − 4 | 9 − 3 | 4 − 1 | 19 − 8 | 6 − 4 | 18 − 7 |
| 19 − 1 | 3 − 2 | 4 − 0 | 5 − 2 | 18 − 3 | 7 − 6 |
| 2 − 0 | 3 − 3 | 19 − 7 | 18 − 1 | 6 − 2 | 7 − 4 |
| 19 − 2 | 4 − 4 | 19 − 8 | 15 − 3 | 6 − 0 | 17 − 2 |
| 2 − 2 | 19 − 5 | 5 − 3 | 6 − 6 | 18 − 5 | 7 − 3 |

# Add and Subtract Within 20



Find each difference.

| | | | | | |
|---|---|---|---|---|---|
| 9 | 3 | 19 | 6 | 12 | 17 |
| − 6 | − 1 | − 7 | − 3 | − 2 | − 6 |
| | | | | | |
| 2 | 4 | 5 | 18 | 7 | 8 |
| − 1 | − 3 | − 1 | − 4 | − 5 | − 6 |
| | | | | | |
| 19 | 4 | 11 | 6 | 17 | 13 |
| − 3 | − 2 | − 0 | − 1 | − 1 | − 2 |
| | | | | | |
| 13 | 19 | 18 | 7 | 7 | 9 |
| − 1 | − 6 | − 2 | − 7 | − 2 | − 8 |
| | | | | | |
| 19 | 14 | 6 | 18 | 17 | 12 |
| − 5 | − 3 | − 5 | − 6 | − 5 | − 1 |

# 64 Add and Subtract Within 20

Use the bar model to answer each question.

① Mrs. Clark picks 7 pink tulips and 5 yellow tulips. How many tulips does she pick all together?

She picks ______ tulips.

| | |
|---|---|
| 7 | 5 |

______

② There are 4 girls and 6 boys at a party. How many children are at the party?

There are ______ children at the party.

| | |
|---|---|
| 4 | 6 |

______

③ Anna has 10 stickers. Then, she gets 7 more stickers. How many stickers does she have now?

Anna has ______ stickers.

| | |
|---|---|
| 10 | 7 |

______

④ There are 4 apples and 4 pears in a fruit bowl. How many pieces of fruit are in the bowl?

There are ______ pieces of fruit in the bowl.

| | |
|---|---|
| 4 | 4 |

______

Use the bar model to answer each question.

1. There are 9 rabbits at a pet store. 6 are brown and the rest are white. How many are white?

   ______ of the rabbits are white.

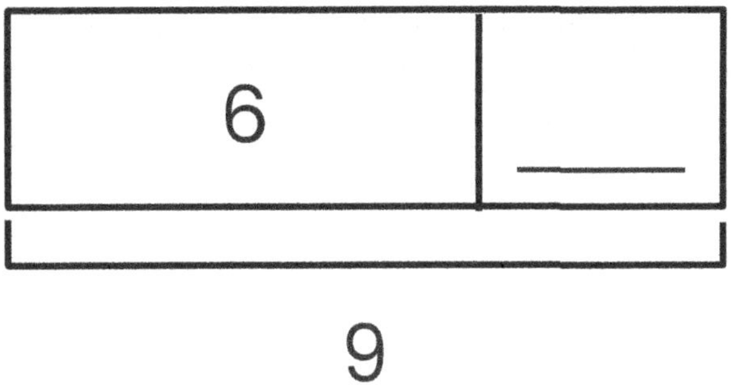


2. Of the 18 kids in a class, 7 wear glasses and the rest do not. How many kids do not wear glasses?

   There are ______ kids who do not wear glasses.

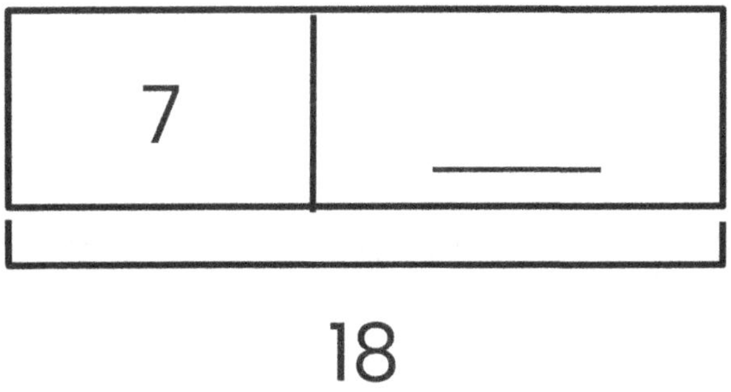


3. David has 12 blocks. 8 are red and the rest are green. How many blocks are green?

   David has ______ green blocks.

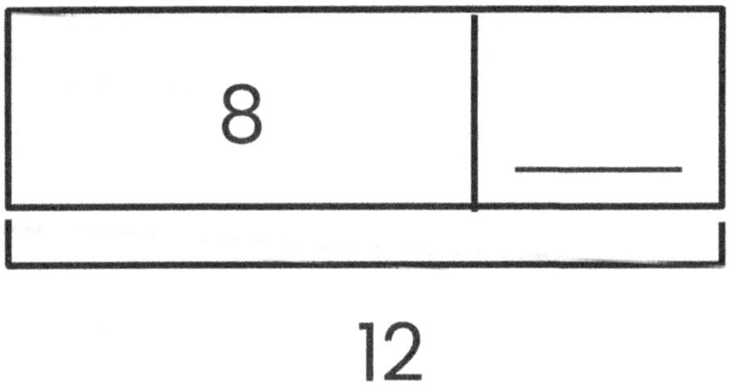


4. There are 16 tomatoes in a garden. 8 are ripe. How many are not ripe?

   There are ______ tomatoes that are not ripe.

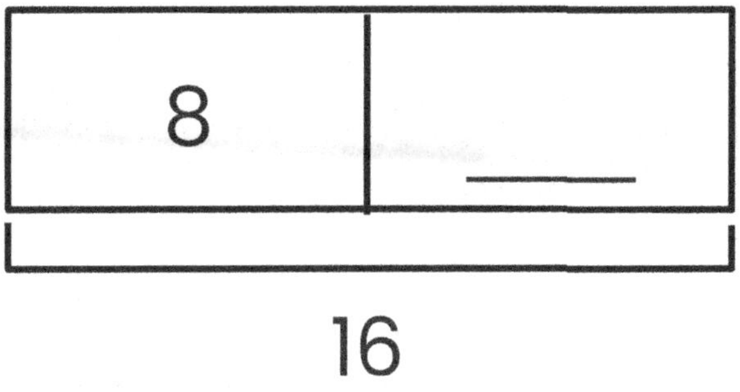

# Add and Subtract Within 20

Use the bar model to show each problem. Write an equation and solve.

1. Samir counts 13 birds in his yard. 6 birds are in a tree. How many birds are not in a tree?

| ____ | ____ |
| --- | --- |

____

____ ☐ ____ = ____

There are ______ birds that are not in a tree.

2. There are 7 short-sleeved shirts and 8 long-sleeved shirts in Cody's closet. How many shirt are in the closet?

| ____ | ____ |
| --- | --- |

____

____ ☐ ____ = ____

There are ______ shirts in the closet.

# Add and Subtract Within 20



Use the bar model to show each problem. Write an equation and solve.

1. Mrs. Jones gives the class 10 choices for a movie to watch. Alex has seen 3 of the movies before. How many has he not seen before?

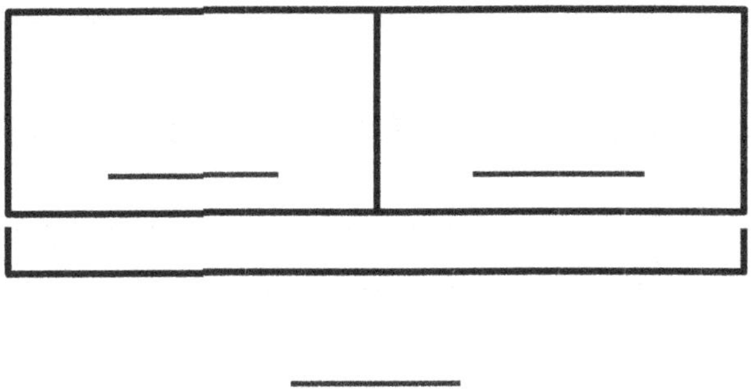

____ □ ____ = ____

Alex has not seen ________ of the movies before.

2. There are 5 black horses and 7 brown horses on a farm. How many horses are on the farm?

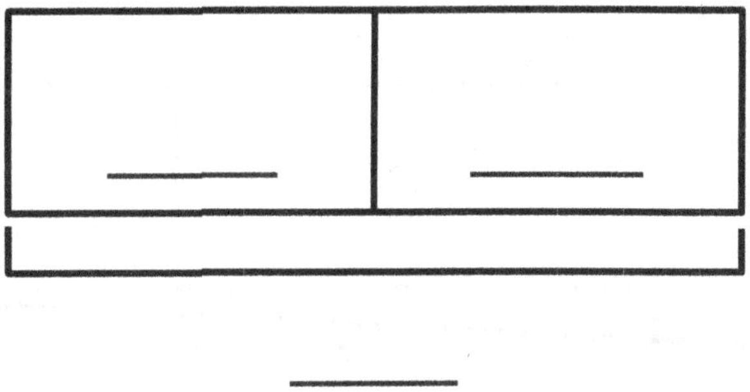

____ □ ____ = ____

There are ________ horses on the farm.

# 68 Add and Subtract Within 20

Write two equations that could describe each bar model.

1
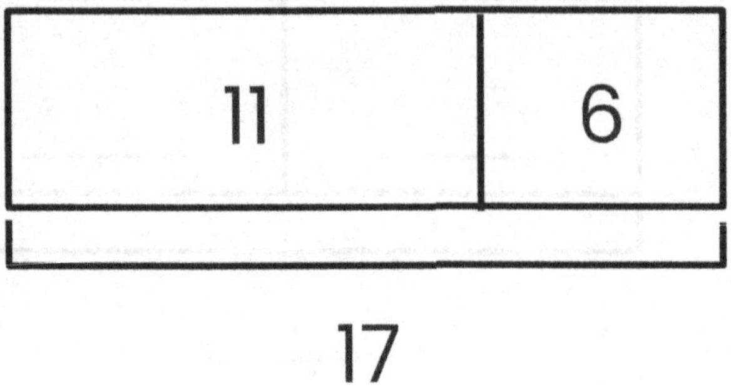


____ + ____ = ____

____ - ____ = ____

2
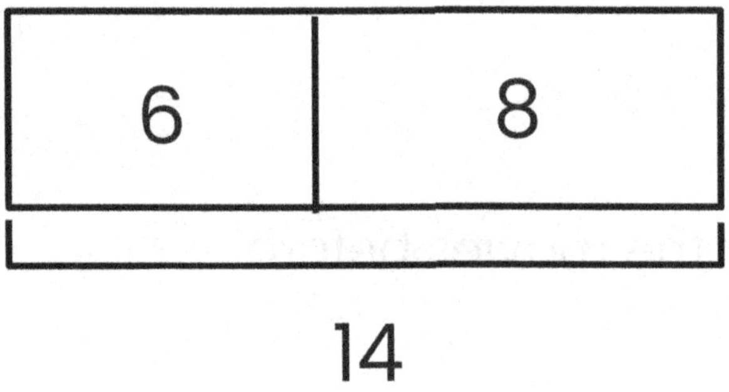


____ + ____ = ____

____ - ____ = ____

3
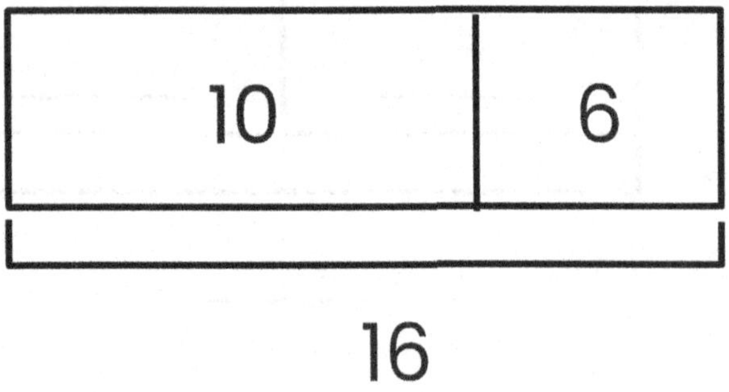


____ + ____ = ____

____ - ____ = ____

4
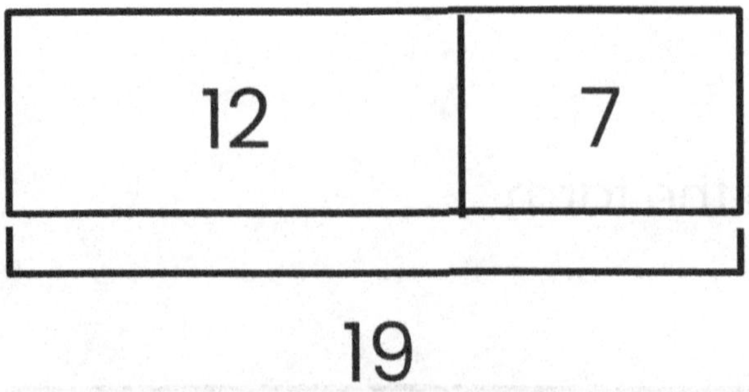


____ + ____ = ____

____ - ____ = ____

# Add and Subtract Within 20



Write the correct symbol (+, -, or = ) to complete each equation.

1. 10 ◯ 7 = 17
2. 19 - 5 ◯ 14
3. 16 ◯ 12 + 4
4. 18 = 8 ◯ 10
5. 8 ◯ 3 = 5
6. 17 - 4 ◯ 13
7. 16 ◯ 1 + 15
8. 6 = 8 ◯ 2
9. 5 ◯ 7 = 12
10. 11 = 9 ◯ 2
11. 14 ◯ 3 = 11
12. 17 + 2 ◯ 19
13. 10 ◯ 10 = 0
14. 13 - 1 ◯ 12

# Add and Subtract Within 20

For each set of three numbers given, write four related addition and subtraction facts.

1

| | | |
|---|---|---|
| 15 | 7 | 8 |

____ + ____ = ____

____ + ____ = ____

____ - ____ = ____

____ - ____ = ____

2

| | | |
|---|---|---|
| 8 | 12 | 4 |

____ + ____ = ____

____ + ____ = ____

____ - ____ = ____

____ - ____ = ____

Circle whether each equation is **TRUE** or **FALSE.**

① **TRUE or FALSE ?**

8 + 3 = 5 + 6

② **TRUE or FALSE ?**

17 - 2 = 15 - 1

③ **TRUE or FALSE ?**

18 - 2 = 9 + 7

④ **TRUE or FALSE ?**

7 + 3 = 1 + 9

⑤ **TRUE or FALSE ?**

14 - 5 = 18 - 9

⑥ **TRUE or FALSE ?**

6 + 7 = 19 - 4

---

Fill in any numbers that would make each equation true.

⑦ 6 + 3 = ___ - ___

⑧ 10 - 5 = ___ + ___

⑨ 12 + 2 = ___ - ___

⑩ 16 - 5 = ___ + ___

⑪ 2 + 11 = ___ - ___

⑫ 8 - 4 = ___ + ___

Draw a picture to help you solve each addition equation.

| | |
|---|---|
| 2 + 3 + 4 = ____ | 7 + 2 + 4 = ____ |
| 3 + 5 + 5 = ____ | 8 + 2 + 1 = ____ |
| 1 + 4 + 9 = ____ | 6 + 2 + 7 = ____ |

Read each word problem. Draw a picture in the box, then fill in the numbers to complete the equation.

Landon has toy cars. 6 are blue, 2 are green, and 5 are red. How many toy cars does he have?

_____ + _____ + _____ = _____

Eric picks 3 green apples, 8 red apples, and 4 yellow apples. How many apples does he pick?

_____ + _____ + _____ = _____

5 people in a group have brown hair, 4 have blonde hair, and 4 have black hair. How many people are in the group?

_____ + _____ + _____ = _____

# Add and Subtract Within 20

Each word problem below takes more than one step to solve. Write two equations to help you solve.

1. Zoe has 8 crackers. She gives 3 of them to Ben. Then, she gets 5 more crackers for herself. How many crackers does Zoe have now?

    ____ ○ ____ = ____

    ____ ○ ____ = ____

    Zoe has ____ crackers.

2. There are 15 vehicles in a parking lot. 3 of them are trucks and the rest are cars. 4 of the cars are red. How many cars are not red?

    ____ ○ ____ = ____

    ____ ○ ____ = ____

    ____ of the cars are not red.

(3) Mae and Beth each pick 7 flowers. Then, Beth drops 2 of her flowers. How many flowers do they have left?

____ ◯ ____ = ____

____ ◯ ____ = ____

Mae and Beth have ______ flowers left.

(4) Derek has 10 trading cards. He gets a new pack of 8 cards. Then, he gives 4 of them to his brother. How many cards foes Derek have left?

____ ◯ ____ = ____

____ ◯ ____ = ____

Derek has ______ cards.

# Equal Groups and Arrays

Write an equation to show each number as the sum of two equal addends.

1. 10 = 5 + 5
2. 8 = ____ + ____
3. 12 = ____ + ____
4. 18 = ____ + ____
5. 6 = ____ + ____
6. 2 = ____ + ____
7. 16 = ____ + ____
8. 20 = ____ + ____
9. 4 = ____ + ____
10. 14 = ____ + ____

Draw to show equal groups. Then write an equation to solve.

Alice has 16 pretzels and divides them into 2 equal groups to share with Dean. How many pretzels are in each group?

____ = ____ + ____

____ pretzels

Draw equal groups to help you solve each problem.

Jane lines up buttons in 3 rows of 5 buttons each. How many buttons are there?

______ buttons

Wesley deals 6 cards to each of the 3 players in a game. How many cards does he deal?

______ cards

Joey puts his toy cars in 4 rows of 4 cars each. How many cars does he have?

______ cars

# 78 Equal Groups and Arrays

Answer the following questions about equal groups.

1. Shelby puts star stickers on a paper. She makes 3 rows with 4 stickers in each row. Which picture shows Shelby's stickers?

○ 

○ 

○ 

2. How many star stickers does Shelby have?

_____ stickers

3. Write an equation to show how many stickers Shelby has.

_____ + _____ + _____ = _____

Write the number of rows and the number of objects in each row. Then, complete the equation to find the total number of objects.

1

______ rows

______ ice cream cones in each row

______ + ______ = ______

2

______ rows

______ basketballs in each row

______ + ______ + ______ = ______

3

______ rows

______ apples in each row

______ + ______ + ______ = ______

# Two-Digit Operations

Use the hundred chart below to help solve each equation.

① 38 + 7 = ____

② 97 - 16 = ____

③ 84 + 12 = ____

④ 47 - 15 = ____

⑤ 55 + 9 = ____

| | | | | | | | | | |
|---|---|---|---|---|---|---|---|---|---|
| 1 | 2 | 3 | 4 | 5 | 6 | 7 | 8 | 9 | 10 |
| 11 | 12 | 13 | 14 | 15 | 16 | 17 | 18 | 19 | 20 |
| 21 | 22 | 23 | 24 | 25 | 26 | 27 | 28 | 29 | 30 |
| 31 | 32 | 33 | 34 | 35 | 36 | 37 | 38 | 39 | 40 |
| 41 | 42 | 43 | 44 | 45 | 46 | 47 | 48 | 49 | 50 |
| 51 | 52 | 53 | 54 | 55 | 56 | 57 | 58 | 59 | 60 |
| 61 | 62 | 63 | 64 | 65 | 66 | 67 | 68 | 69 | 70 |
| 71 | 72 | 73 | 74 | 75 | 76 | 77 | 78 | 79 | 80 |
| 81 | 82 | 83 | 84 | 85 | 86 | 87 | 88 | 89 | 90 |
| 91 | 92 | 93 | 94 | 95 | 96 | 97 | 98 | 99 | 100 |

Use the hundred chart above to help answer each question.

⑥ Betty has a box of 48 crayons. Taylor has a box of 24 crayons. How many crayons do they have all together?

____ + ____ = ____

⑦ There are 32 slices of pizza at a birthday party. 25 of them get eaten. How many slices of pizza are left over?

____ - ____ = ____

Use the number line below to help solve each equation.

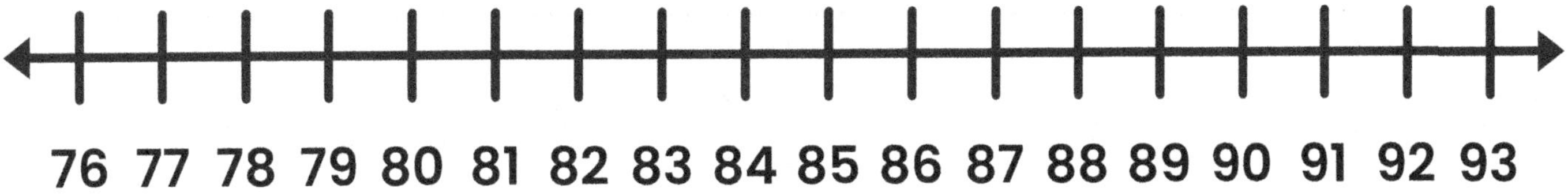


① 82 + 7 = ____

② 88 - 11 = ____

③ 89 - 5 = ____

④ 77 + 12 = ____

⑤ 91 - 6 = ____

⑥ 79 + 9 = ____

---

Use the number line below to help answer each question.

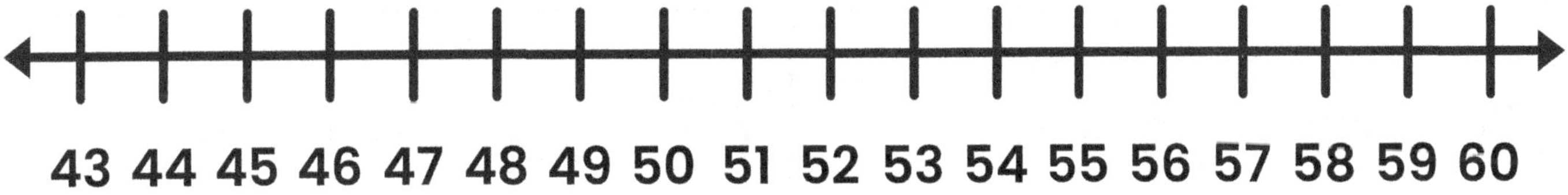


⑦ Oliver is 46 inches tall. His brother Marcus is 7 inches taller than him. How tall is Marcus?

_____ + _____ = _____

⑧ Christa is watching a movie that is 60 minutes long. She has already watched 48 minutes. How many minutes are left?

_____ - _____ = _____

Re-write each equation vertically as tens and ones to add.

(EX) 34 + 23 = ?

| | Tens | Ones |
|---|---|---|
| | 3 | 4 |
| + | 2 | 3 |
| | 5 | 7 |

5 tens + 7 ones

50 + 7 = 57

(1) 62 + 14 = ?

| | Tens | Ones |
|---|---|---|
| | | |
| + | | |
| | | |

____ tens + ____ ones

____ + ____ = ____

(2) 51 + 38 = ?

| | Tens | Ones |
|---|---|---|
| | | |
| + | | |
| | | |

____ tens + ____ ones

____ + ____ = ____

(3) 45 + 32 = ?

| | Tens | Ones |
|---|---|---|
| | | |
| + | | |
| | | |

____ tens + ____ ones

____ + ____ = ____

Find each sum.

| | | | | |
|---|---|---|---|---|
| 65 + 33 = 98 | 40 + 19 | 72 + 13 | 35 + 44 | 18 + 50 |
| 22 + 17 | 56 + 43 | 43 + 35 | 22 + 46 | 70 + 19 |
| 41 + 48 | 74 + 23 | 57 + 10 | 35 + 32 | 16 + 22 |
| 20 + 50 | 45 + 12 | 51 + 17 | 24 + 33 | 14 + 52 |

# Two-Digit Operations

Re-write each equation vertically as tens and ones to subtract.

**EX** 87 - 26 = ?

| | Tens | Ones |
|---|---|---|
| | 8 | 7 |
| - | 2 | 6 |
| | 6 | 1 |

6 tens + 1 ones

60 + 1 = 61

**1** 95 - 31 = ?

| | Tens | Ones |
|---|---|---|
| | | |
| - | | |
| | | |

____ tens + ____ ones

____ + ____ = ____

**2** 49 - 34 = ?

| | Tens | Ones |
|---|---|---|
| | | |
| - | | |
| | | |

____ tens + ____ ones

____ + ____ = ____

**3** 68 - 15 = ?

| | Tens | Ones |
|---|---|---|
| | | |
| - | | |
| | | |

____ tens + ____ ones

____ + ____ = ____

Find each difference.

| | | | | |
|---|---|---|---|---|
| $95 - 23 = 72$ | $89 - 16$ | $78 - 63$ | $60 - 40$ | $54 - 12$ |
| $49 - 27$ | $62 - 51$ | $58 - 17$ | $79 - 26$ | $82 - 72$ |
| $68 - 15$ | $74 - 43$ | $99 - 74$ | $43 - 30$ | $56 - 54$ |
| $85 - 34$ | $47 - 22$ | $39 - 33$ | $76 - 50$ | $58 - 24$ |

Draw tens and ones to represent the addition problem. Circle to show how you can make a new group of ten. Then, complete the equation and solve.

EX

27 + 5 = ?

27 + 3 + 2 = ?

30 + 2 = 32

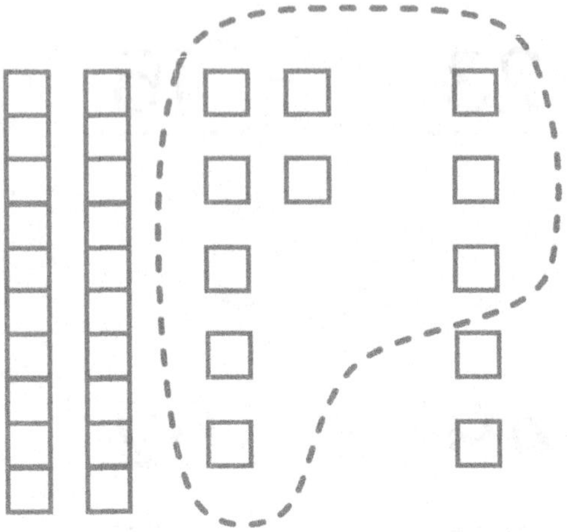

27 + 5 = 32

1

56 + 8 = ?

____ + ____ + ____ = ?

____ + ____ = ____

56 + 8 = ____

(2)

$49 + 7 = ?$

____ + ____ + ____ = ?

____ + ____ = ____

49 + 7 = ____

(3)

$63 + 9 = ?$

____ + ____ + ____ = ?

____ + ____ = ____

63 + 9 = ____

Use the strategy of making a ten to solve each addition problem.

**EX**

36 + 7 = ?

36 + 4 + 3 = ?

40 + 3 = 43

36 + 7 = 43

**1**

59 + 3 = ?

____ + ____ + ____ = ?

____ + ____ = ____

59 + 3 = ____

**2**

87 + 8 = ?

____ + ____ + ____ = ?

____ + ____ = ____

87 + 8 = ____

**3**

76 + 6 = ?

____ + ____ + ____ = ?

____ + ____ = ____

76 + 6 = ____

(4)

65 + 9 = ?

____ + ____ + ____ = ?

____ + ____ = ____

65 + 9 = ____

(5)

78 + 6 = ?

____ + ____ + ____ = ?

____ + ____ = ____

78 + 6 = ____

(6)

19 + 4 = ?

____ + ____ + ____ = ?

____ + ____ = ____

19 + 4 = ____

(7)

37 + 5 = ?

____ + ____ + ____ = ?

____ + ____ = ____

37 + 5 = ____

Solve each equation by adding tens and ones. Regroup if needed by making a ten with the ones and adding it to the tens column.

(EX) 28 + 53 = ?

| | Tens | Ones |
|---|---|---|
| | 1 | |
| | 2 | 8 |
| + | 5 | 3 |
| | 8 | 1 |

28 + 53 = 81

(1) 37 + 45 = ?

| | Tens | Ones |
|---|---|---|
| | ☐ | |
| | 3 | 7 |
| + | 4 | 5 |
| | | |

37 + 45 = ____

(2) 76 + 14 = ?

| | Tens | Ones |
|---|---|---|
| | ☐ | |
| | 7 | 6 |
| + | 1 | 4 |
| | | |

76 + 14 = ____

(3) 13 + 29 = ?

| | Tens | Ones |
|---|---|---|
| | ☐ | |
| | 1 | 3 |
| + | 2 | 9 |
| | | |

13 + 29 = ____

(4) 57 + 23 = ?

| Tens | Ones |
|---|---|
| ☐ | |
| 5 | 7 |
| + 2 | 3 |
| | |

57 + 23 = ____

(5) 26 + 65 = ?

| Tens | Ones |
|---|---|
| ☐ | |
| 2 | 6 |
| + 6 | 5 |
| | |

26 + 65 = ____

(6) 19 + 15 = ?

| Tens | Ones |
|---|---|
| ☐ | |
| 1 | 9 |
| + 1 | 5 |
| | |

19 + 15 = ____

(7) 44 + 38 = ?

| Tens | Ones |
|---|---|
| ☐ | |
| 4 | 4 |
| + 3 | 8 |
| | |

44 + 38 = ____

Re-write each equation vertically as tens and ones to add. Regroup if needed by making a ten with the ones and adding it to the tens column.

(EX) 48 + 34 = ?

| | Tens | Ones |
|---|---|---|
| | 1 | |
| | 4 | 8 |
| + | 3 | 4 |
| | 8 | 2 |

48 + 34 = 82

(1) 27 + 14 = ?

| | Tens | Ones |
|---|---|---|
| | | |
| | | |
| + | | |
| | | |

27 + 14 = ____

(2) 16 + 45 = ?

| | Tens | Ones |
|---|---|---|
| | | |
| | | |
| + | | |
| | | |

16 + 45 = ____

(3) 39 + 36 = ?

| | Tens | Ones |
|---|---|---|
| | | |
| | | |
| + | | |
| | | |

39 + 36 = ____

Find each sum. Regroup if needed.

| | | | | |
|---|---|---|---|---|
| 65 | 43 | 62 | 45 | 29 |
| + 26 | + 39 | + 19 | + 47 | + 51 |
| 91 | | | | |
| 35 | 56 | 47 | 28 | 72 |
| + 17 | + 28 | + 35 | + 46 | + 19 |
| 42 | 34 | 59 | 35 | 16 |
| + 48 | + 28 | + 17 | + 36 | + 26 |
| 25 | 45 | 52 | 44 | 25 |
| + 55 | + 17 | + 29 | + 16 | + 56 |

Add each set of three numbers. Regroup if needed by making a ten with the ones and adding it to the tens column.

EX

| | Tens | Ones |
|---|---|---|
| | 1 | |
| | 2 | 8 |
| | 3 | 1 |
| + | 3 | 3 |
| | 9 | 2 |

1

| | Tens | Ones |
|---|---|---|
| | 3 | 2 |
| | 1 | 9 |
| + | 4 | 5 |

2

| | Tens | Ones |
|---|---|---|
| | 1 | 7 |
| | 2 | 1 |
| + | 5 | 4 |

3

| | Tens | Ones |
|---|---|---|
| | 6 | 0 |
| | 1 | 7 |
| + | 1 | 8 |

4

$$\begin{array}{r} 24 \\ 32 \\ +\ 16 \\ \hline \end{array}$$

5

$$\begin{array}{r} 48 \\ 11 \\ +\ 23 \\ \hline \end{array}$$

6

$$\begin{array}{r} 25 \\ 41 \\ +\ 17 \\ \hline \end{array}$$

Add each set of four numbers. Regroup if needed by making a ten with the ones and adding it to the tens column.

EX

| Tens | Ones |
|---|---|
| 1 | |
| 1 | 3 |
| 3 | 4 |
| 2 | 7 |
| + 2 | 1 |
| 9 | 5 |

1

| Tens | Ones |
|---|---|
| ☐ | |
| 3 | 1 |
| 2 | 9 |
| 1 | 4 |
| + 1 | 5 |
| | |

2

| Tens | Ones |
|---|---|
| ☐ | |
| 2 | 2 |
| 3 | 8 |
| 1 | 2 |
| + 2 | 6 |
| | |

3

| Tens | Ones |
|---|---|
| ☐ | |
| 1 | 1 |
| 1 | 8 |
| 4 | 2 |
| + 2 | 6 |
| | |

4

  24
  27
  12
+ 16

5

  14
  23
  18
+ 25

6

  13
  26
  35
+ 14

# Two-Digit Operations

Solve each equation by subtracting tens and ones. Regroup if needed by borrowing a ten from the tens column and adding it to the ones column.

(EX) 93 - 55 = ?

| | Tens | Ones |
|---|---|---|
| | 8 | 13 |
| | ~~9~~ | ~~3~~ |
| - | 5 | 5 |
| | 3 | 8 |

93 - 55 = 81

(1) 41 - 15 = ?

| | Tens | Ones |
|---|---|---|
| | ☐ | ☐ |
| | 4 | 1 |
| - | 1 | 5 |
| | | |

41 - 15 = ____

(2) 73 - 46 = ?

| | Tens | Ones |
|---|---|---|
| | ☐ | ☐ |
| | 7 | 3 |
| - | 4 | 6 |
| | | |

73 - 46 = ____

(3) 53 - 27 = ?

| | Tens | Ones |
|---|---|---|
| | ☐ | ☐ |
| | 5 | 3 |
| - | 2 | 7 |
| | | |

53 - 27 = ____

④ 80 - 23 = ?

| Tens | Ones |
| --- | --- |
| ☐ | ☐ |
| 8 | 0 |
| - 2 | 3 |
| | |

80 - 23 = ____

⑤ 66 - 27 = ?

| Tens | Ones |
| --- | --- |
| ☐ | ☐ |
| 6 | 6 |
| - 2 | 7 |
| | |

66 - 27 = ____

⑥ 32 - 18 = ?

| Tens | Ones |
| --- | --- |
| ☐ | ☐ |
| 3 | 2 |
| - 1 | 8 |
| | |

32 - 18 = ____

⑦ 54 - 28 = ?

| Tens | Ones |
| --- | --- |
| ☐ | ☐ |
| 5 | 4 |
| - 2 | 8 |
| | |

54 - 28 = ____

Re-write each equation as tens and ones to subtract. Regroup if needed by borrowing a ten from the tens column and adding it to the ones column.

**EX** 42 - 13 = ?

| | Tens | Ones |
|---|---|---|
| | 3 | 12 |
| | ~~4~~ | ~~2~~ |
| − | 2 | 9 |
| | 1 | 3 |

42 - 13 = 13

**1** 87 - 19 = ?

| | Tens | Ones |
|---|---|---|
| | □ | □ |
| | | |
| − | | |
| | | |

87 - 19 = ____

**2** 61 - 34 = ?

| | Tens | Ones |
|---|---|---|
| | □ | □ |
| | | |
| − | | |
| | | |

61 - 34 = ____

**3** 50 - 27 = ?

| | Tens | Ones |
|---|---|---|
| | □ | □ |
| | | |
| − | | |
| | | |

50 - 27 = ____

Find each difference. Regroup if needed.

| | | | | |
|---|---|---|---|---|
| 5 15<br>65<br>− 26<br>39 | 43<br>− 19 | 82<br>− 37 | 45<br>− 27 | 70<br>− 51 |
| 35<br>− 18 | 56<br>− 38 | 47<br>− 29 | 63<br>− 46 | 72<br>− 49 |
| 50<br>− 22 | 64<br>− 38 | 85<br>− 67 | 34<br>− 16 | 76<br>− 27 |
| 92<br>− 54 | 75<br>− 17 | 64<br>− 29 | 54<br>− 37 | 82<br>− 43 |

# Three-Digit Operations

Add and subtract to find the numbers that are 10 less and 10 more than each number shown.

| 10 Less | | 10 More |
|---|---|---|
| 202 | **212** | 222 |

| 10 Less | | 10 More |
|---|---|---|
| | **334** | |

| 10 Less | | 10 More |
|---|---|---|
| | **829** | |

| 10 Less | | 10 More |
|---|---|---|
| | **163** | |

| 10 Less | | 10 More |
|---|---|---|
| | **642** | |

| 10 Less | | 10 More |
|---|---|---|
| | **918** | |

| 10 Less | | 10 More |
|---|---|---|
| | **781** | |

| 10 Less | | 10 More |
|---|---|---|
| | **475** | |

| 10 Less | | 10 More |
|---|---|---|
| | **597** | |

| 10 Less | | 10 More |
|---|---|---|
| | **156** | |

Add and subtract to find the numbers that are 100 less and 100 more than each number shown.

| 100 Less | | 100 More |
|---|---|---|
| 705 | **805** | 905 |

| 100 Less | | 100 More |
|---|---|---|
| | **298** | |

| 100 Less | | 100 More |
|---|---|---|
| | **347** | |

| 100 Less | | 100 More |
|---|---|---|
| | **762** | |

| 100 Less | | 100 More |
|---|---|---|
| | **481** | |

| 100 Less | | 100 More |
|---|---|---|
| | **129** | |

| 100 Less | | 100 More |
|---|---|---|
| | **650** | |

| 100 Less | | 100 More |
|---|---|---|
| | **573** | |

| 100 Less | | 100 More |
|---|---|---|
| | **714** | |

| 100 Less | | 100 More |
|---|---|---|
| | **836** | |

# Three-Digit Operations

Re-write each equation vertically as hundreds, tens, and ones to add.

① $421 + 536 = ?$

| | Hundreds | Tens | Ones |
|---|---|---|---|
| + | | | |
| | | | |

_____ hundreds + _____ tens + _____ ones

_______ + _______ + _______ = _______

② $142 + 326 = ?$

| | Hundreds | Tens | Ones |
|---|---|---|---|
| + | | | |
| | | | |

_____ hundreds + _____ tens + _____ ones

_______ + _______ + _______ = _______

Find each sum.

| | |
|---|---|
| ① <br> 107 <br> + 642 | ② <br> 223 <br> + 571 |
| ③ <br> 710 <br> + 168 | ④ <br> 392 <br> + 406 |

⑤ 422 + 257 = ____

⑥ 650 + 335 = ____

⑦ 232 + 505 = ____

# Three-Digit Operations

Re-write each equation vertically as hundreds, tens, and ones to subtract.

① $679 - 238 = ?$

| | Hundreds | Tens | Ones |
|---|---|---|---|
| | | | |
| − | | | |
| | | | |

_____ hundreds + _____ tens + _____ ones

______ + ______ + ______ = ______

② $583 - 182 = ?$

| | Hundreds | Tens | Ones |
|---|---|---|---|
| | | | |
| − | | | |
| | | | |

_____ hundreds + _____ tens + _____ ones

______ + ______ + ______ = ______

Find each difference.

① 
$$\begin{array}{r} 957 \\ -\ 723 \\ \hline \end{array}$$

② 
$$\begin{array}{r} 764 \\ -\ 513 \\ \hline \end{array}$$

③ 
$$\begin{array}{r} 782 \\ -\ 461 \\ \hline \end{array}$$

④ 
$$\begin{array}{r} 357 \\ -\ 202 \\ \hline \end{array}$$

⑤ 
$$\begin{array}{r} 653 \\ -\ 242 \\ \hline \end{array}$$

⑥ 
$$\begin{array}{r} 987 \\ -\ 456 \\ \hline \end{array}$$

⑦ 
$$\begin{array}{r} 479 \\ -\ 307 \\ \hline \end{array}$$

# Three-Digit Operations

Solve by adding tens, hundreds, and ones. Regroup if needed by making a ten with the ones and adding it to the tens column or by making a hundred with the tens and adding it to the hundreds column.

① 265 + 387 = ?

| | Hundreds | Tens | Ones |
|---|---|---|---|
| | □ | □ | |
| | 2 | 6 | 5 |
| + | 3 | 8 | 7 |
| | | | |

265 + 387 = _____

② 452 + 109 = ?

| | Hundreds | Tens | Ones |
|---|---|---|---|
| | □ | □ | |
| | 4 | 5 | 2 |
| + | 1 | 0 | 9 |
| | | | |

452 + 109 = _____

Re-write each equation vertically as hundreds, tens, and ones to add. Regroup if needed.

① $347 + 284 = ?$

| | Hundreds | Tens | Ones |
|---|---|---|---|
| | ☐ | ☐ | |
| + | | | |
| | | | |

$347 + 284 =$ _____

② $175 + 548 = ?$

| | Hundreds | Tens | Ones |
|---|---|---|---|
| | ☐ | ☐ | |
| + | | | |
| | | | |

$175 + 548 =$ _____

# Three-Digit Operations

Solve by subtracting tens, hundreds, and ones. Regroup if needed.

(EX) 855 - 381 = ?

| | Hundreds | Tens | Ones |
|---|---|---|---|
| | 7 | 15 | |
| | ~~8~~ | ~~5~~ | 5 |
| - | 3 | 8 | 1 |
| | 4 | 7 | 4 |

855 - 381 = 474

(1) 672 - 135 = ?

| | Hundreds | Tens | Ones |
|---|---|---|---|
| | | | |
| | 6 | 7 | 2 |
| - | 1 | 3 | 5 |
| | | | |

672 - 135 = ______

(2) 408 - 245 = ?

| | Hundreds | Tens | Ones |
|---|---|---|---|
| | ☐ | ☐ | ☐ |
| | 4 | 0 | 8 |
| - | 2 | 4 | 5 |
| | | | |

408 - 245 = ____

(3) 937 - 512 = ?

| | Hundreds | Tens | Ones |
|---|---|---|---|
| | ☐ | ☐ | ☐ |
| | 9 | 3 | 7 |
| - | 5 | 1 | 2 |
| | | | |

937 - 512 = ____

# Three-Digit Operations

Re-write each equation vertically as hundreds, tens, and ones to subtract. Regroup if needed.

EX

629 - 347 = ?

| | Hundreds | Tens | Ones |
|---|---|---|---|
| | 5 | 12 | |
| | ~~6~~ | ~~2~~ | 9 |
| - | 3 | 4 | 7 |
| | 2 | 8 | 2 |

629 - 347 = 282

1

491 - 125 = ?

| | Hundreds | Tens | Ones |
|---|---|---|---|
| | | | |
| | | | |
| - | | | |
| | | | |

491 - 125 = ____

(2) 715 - 225 = ?

| Hundreds | Tens | Ones |
| --- | --- | --- |
| ☐ | ☐ | ☐ |
| − | | |
| | | |

715 - 225 = ____

(3) 948 - 568 = ?

| Hundreds | Tens | Ones |
| --- | --- | --- |
| ☐ | ☐ | ☐ |
| − | | |
| | | |

948 - 568 = ____

Find each sum. Regroup if needed.

①

$$\begin{array}{r} 156 \\ +\ 537 \\ \hline \end{array}$$

②

$$\begin{array}{r} 279 \\ +\ 341 \\ \hline \end{array}$$

③

$$\begin{array}{r} 619 \\ +\ 158 \\ \hline \end{array}$$

④

$$\begin{array}{r} 395 \\ +\ 245 \\ \hline \end{array}$$

⑤

$$\begin{array}{r} 812 \\ +\ 179 \\ \hline \end{array}$$

⑥

$$\begin{array}{r} 276 \\ +\ 584 \\ \hline \end{array}$$

⑦

$$\begin{array}{r} 340 \\ +\ 475 \\ \hline \end{array}$$

Find each difference. Regroup if needed.

1.
```
  8 5 2
- 7 1 3
-------
```

2.
```
  7 6 4
- 3 9 2
-------
```

3.
```
  9 8 1
- 4 6 3
-------
```

4.
```
  6 0 5
- 2 8 2
-------
```

5.
```
  435
- 183
-----
```

6.
```
  308
- 154
-----
```

7.
```
  582
- 406
-----
```

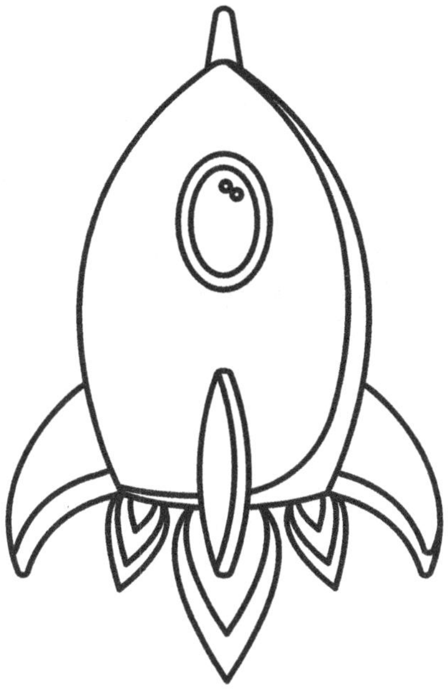

# Geometry

# Two-Dimensional Shapes

Draw a line to match each 2-D shape to its name.

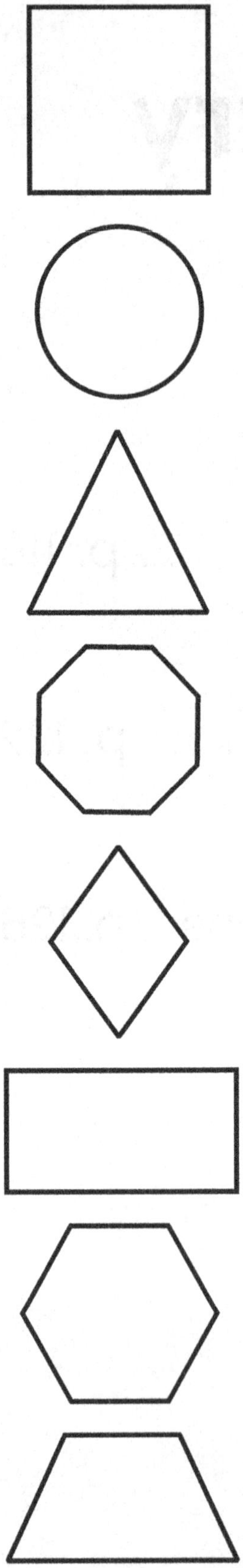

DIAMOND

HEXAGON

RECTANGLE

OCTAGON

TRAPEZOID

TRIANGLE

SQUARE

CIRCLE

Label the parts of the shape using the words from the word bank.

**WORD BANK**

side angle vertex

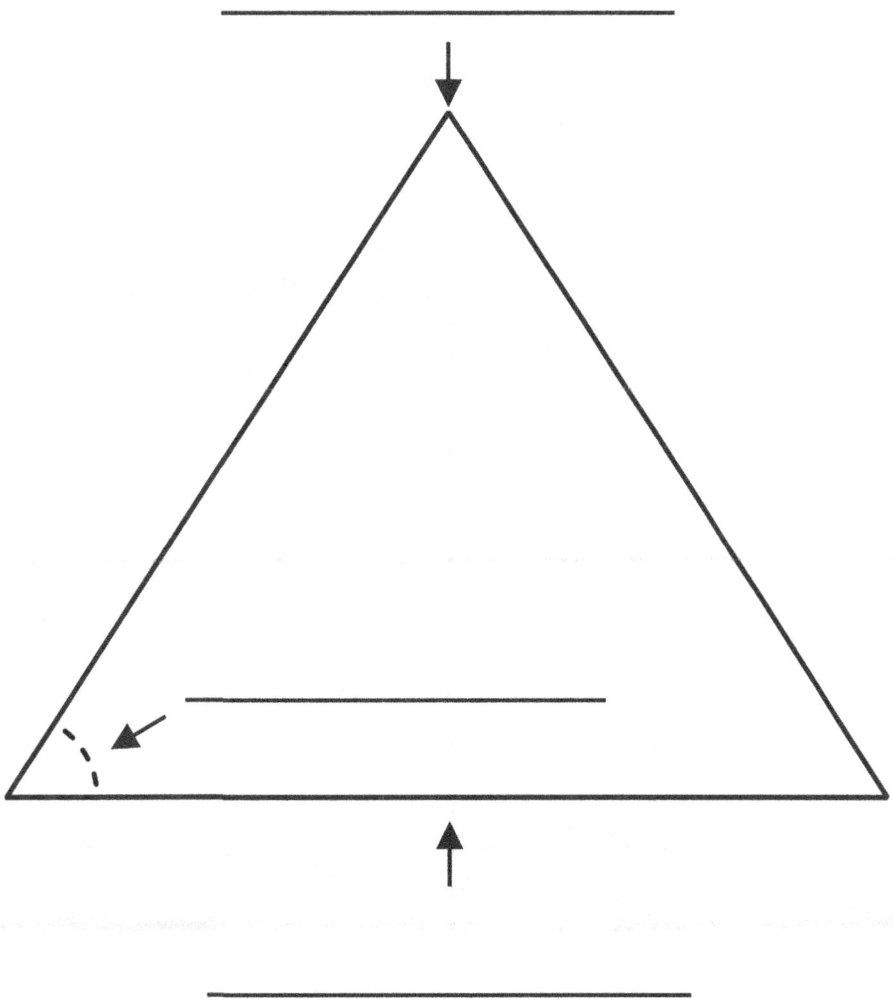

# Two-Dimensional Shapes

Draw each shape. Fill in the blanks with the correct numbers.

| | |
|---|---|
| **SQUARE**<br>______ sides<br>______ vertices<br>______ angles | **TRIANGLE**<br>______ sides<br>______ vertices<br>______ angles |
| **RECTANGLE**<br>______ sides<br>______ vertices<br>______ angles | **HEXAGON**<br>______ sides<br>______ vertices<br>______ angles |
| **PENTAGON**<br>______ sides<br>______ vertices<br>______ angles | **QUADRILATERAL**<br>______ sides<br>______ vertices<br>______ angles |
| **TRAPEZOID**<br>______ sides<br>______ vertices<br>______ angles | **OCTAGON**<br>______ sides<br>______ vertices<br>______ angles |

Fill the in correct bubble to answer each question.

Which of these shapes is NOT a type of quadrilateral?

○ square ○ pentagon ○ diamond

Which of these shapes has more vertices than a rectangle?

○ trapezoid ○ triangle ○ hexagon

Which of these shapes has the fewest number of sides?

○ triangle ○ rectangle ○ pentagon

Which of these shapes has an odd number of sides?

○ hexagon ○ quadrilateral ○ pentagon

Which of these shapes has the greatest number of vertices?

○ octagon ○ hexagon ○ trapezoid

# Two-Dimensional Shapes

In each row, circle the shapes that have the same name. Write the name on the line.

______________________

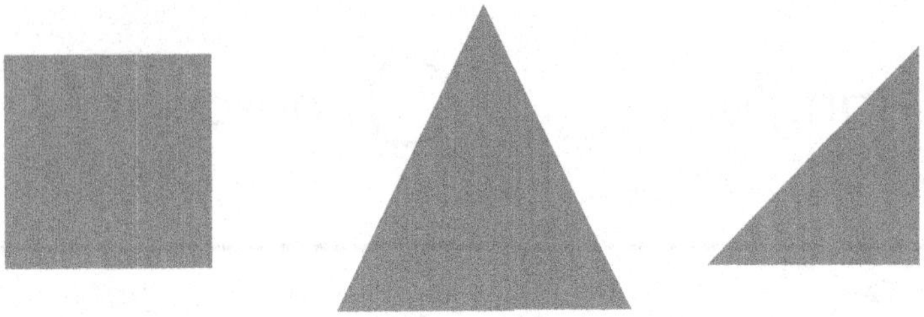

______________________

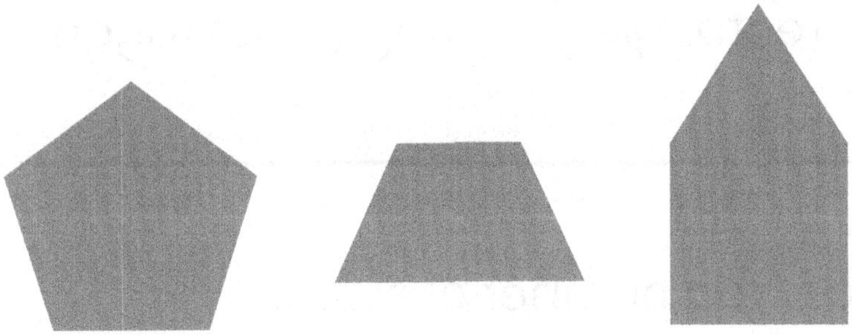

______________________

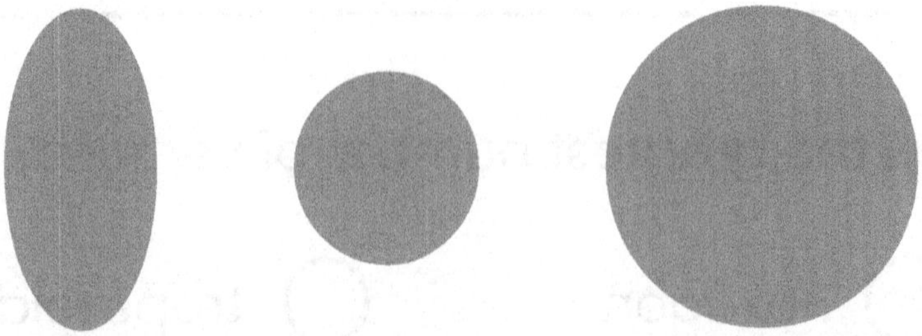

______________________

n each box, draw two different shapes that match the description.

| | |
|---|---|
| Shapes that have 4 sides | Shapes that have more than 4 vertices |
| Shapes that have fewer than 5 angles | Shapes that are curved |

# 122 Three-Dimensional Shapes

Draw a line to match each 3-D shape to its name.

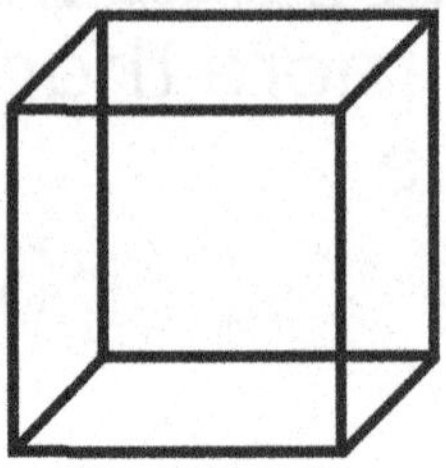

RECTANGULAR PRISM

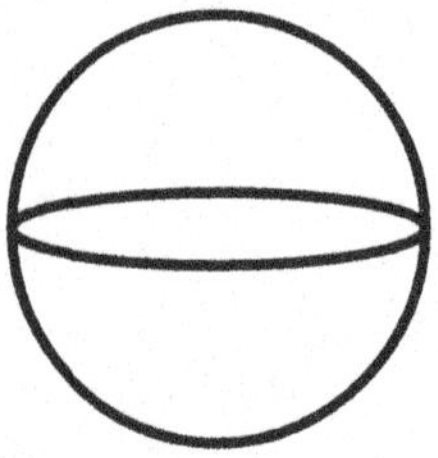

CONE

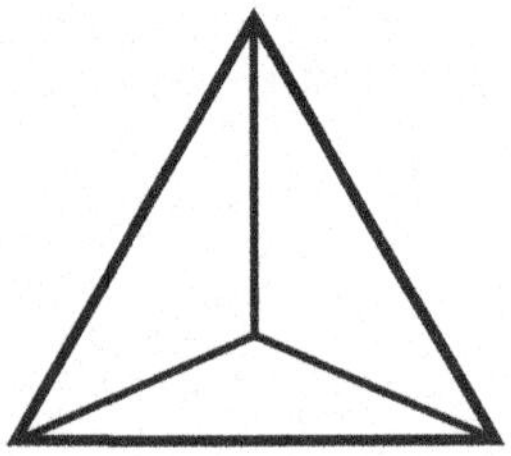

SPHERE

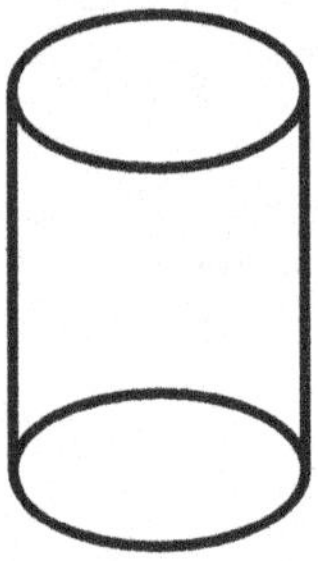

PYRAMID

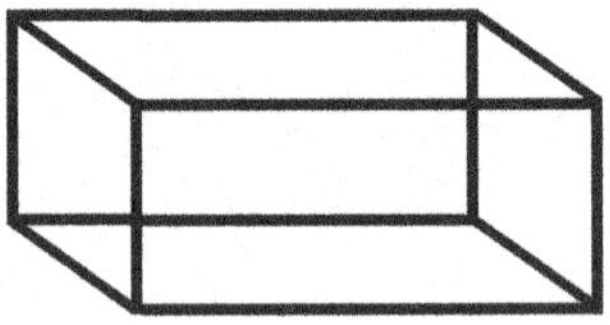

CUBE

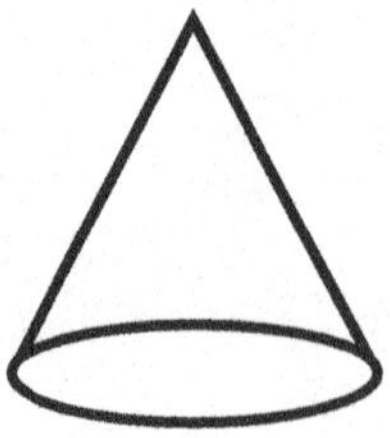

CYLINDER

Label the parts of the three-dimensional shape using the words from the word bank.

**WORD BANK**

edge face vertex

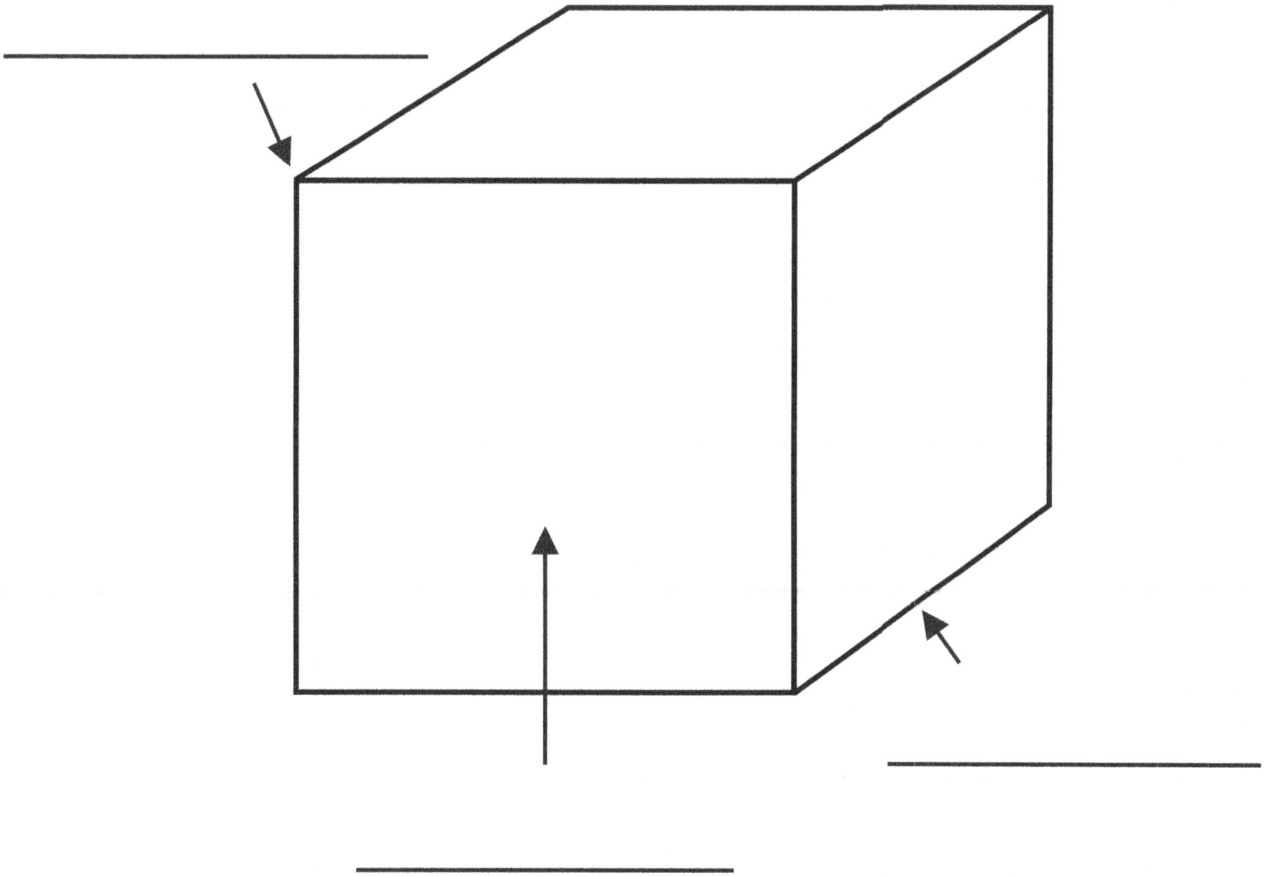

# Three-Dimensional Shapes

Fill the in correct bubble to answer each question.

How many faces does a rectangular prism have?

- ○ 4
- ○ 6
- ○ 8

What shape is each face of a cube?

- ○ triangle
- ○ circle
- ○ square

How many flat faces does a cylinder have?

- ○ 0
- ○ 1
- ○ 2

How many flat faces does a cone have?

- ○ 0
- ○ 1
- ○ 2

Which of these shapes has no vertices?

- ○ sphere
- ○ cube
- ○ pyramid

Circle the 3-D shapes in each group that match the description given.

Circle the solid figures that have **8 vertices**.

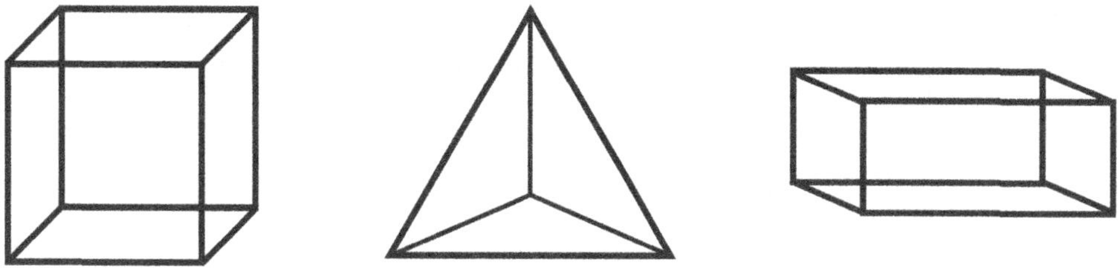

Circle the solid figures that have a **curved surface**.

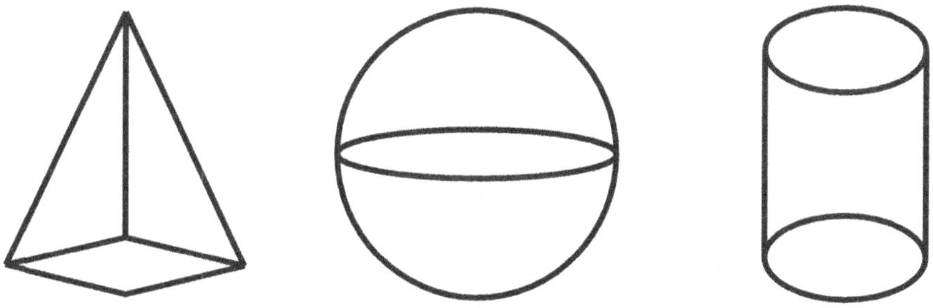

Circle the solid figures that have a **circular base**.

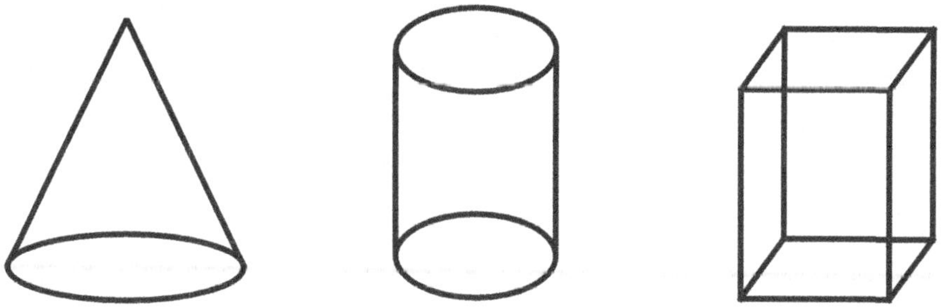

Circle the solid figures that have **fewer than 6 faces.**

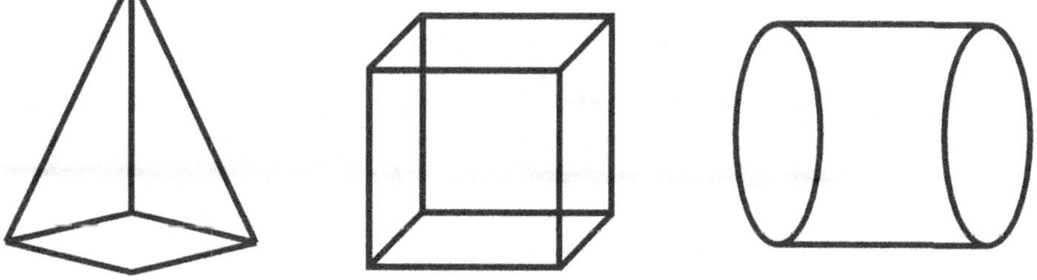

# Shapes to Represent Fractions

Write the number of equal shares shown in each shape.

EX

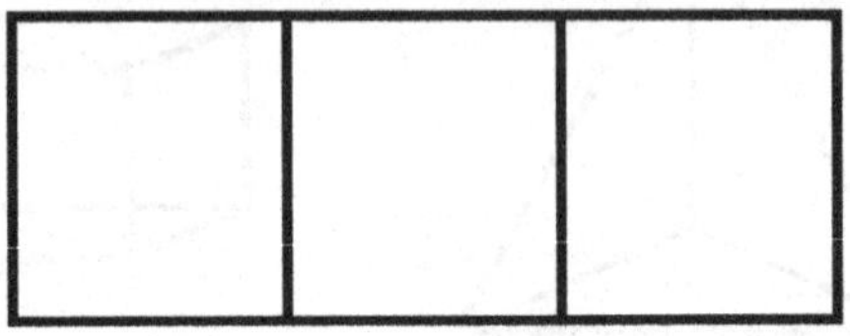

3
______

1

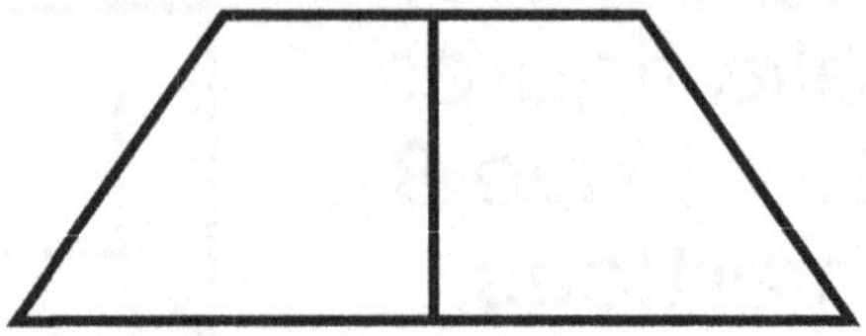

______

2

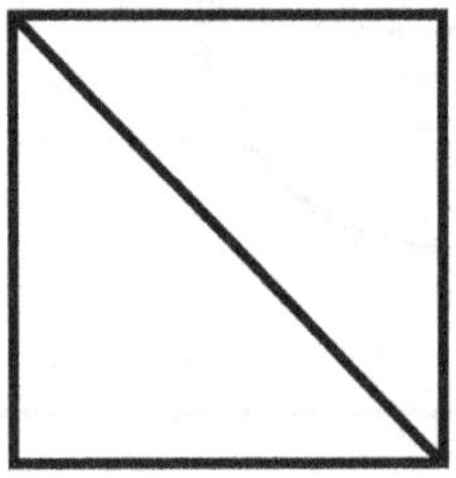

______

3

______

4

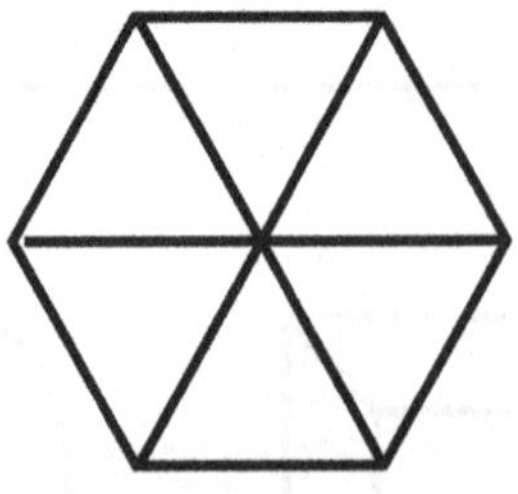

______

5

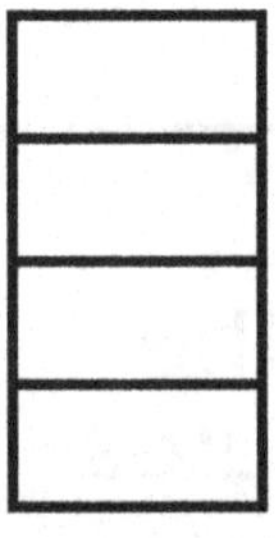

______

# Shapes to Represent Fractions



Write the number of rows, columns, and square tiles within each rectangle.

EX

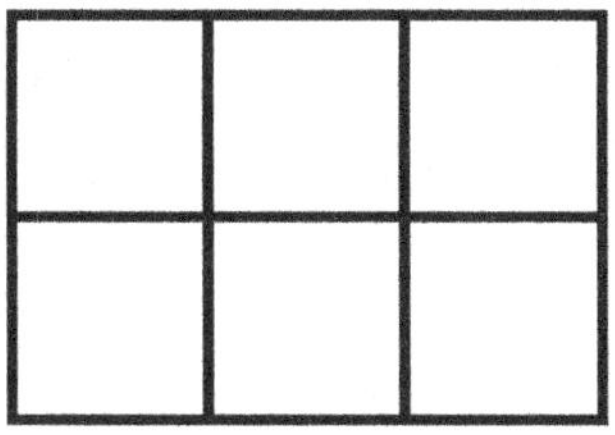

__2__ rows

__3__ columns

__6__ squares

1

______ rows

______ columns

______ squares

2

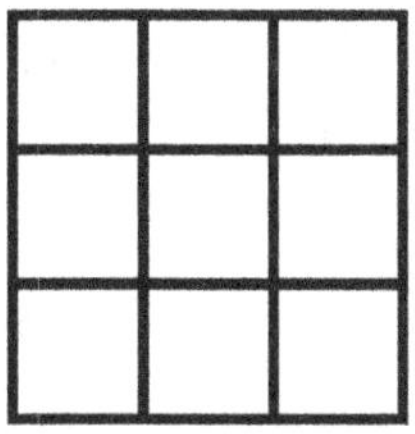

______ rows

______ columns

______ squares

3

______ rows

______ columns

______ squares

4

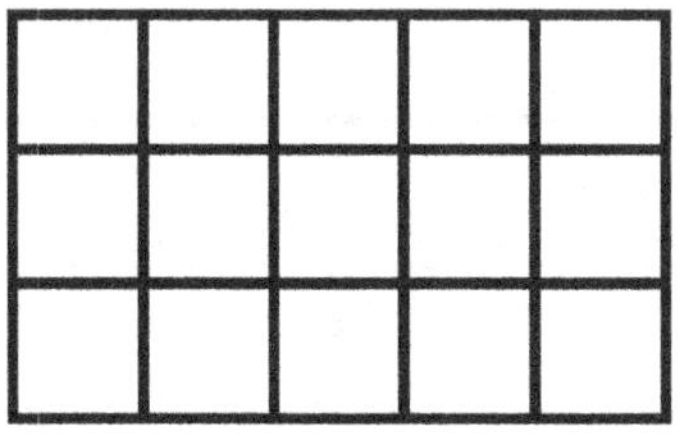

______ rows

______ columns

______ squares

5

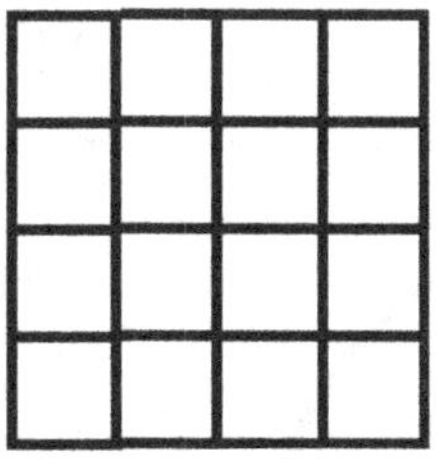

______ rows

______ columns

______ squares

# Shapes to Represent Fractions

Circle whether each shape shows halves, thirds, or fourths.

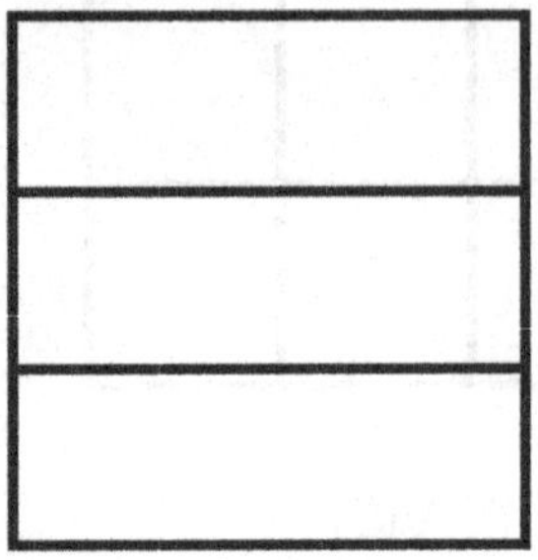

halves

thirds

quarters

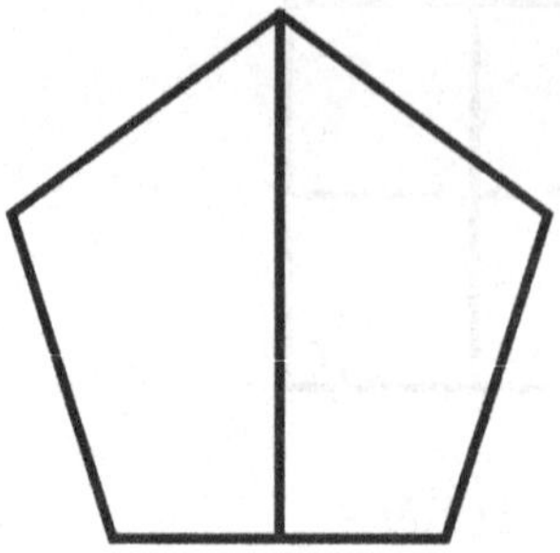

halves

thirds

quarters

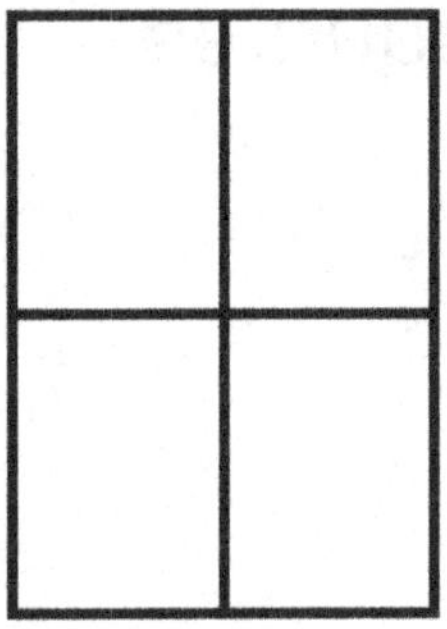

halves

thirds

quarters

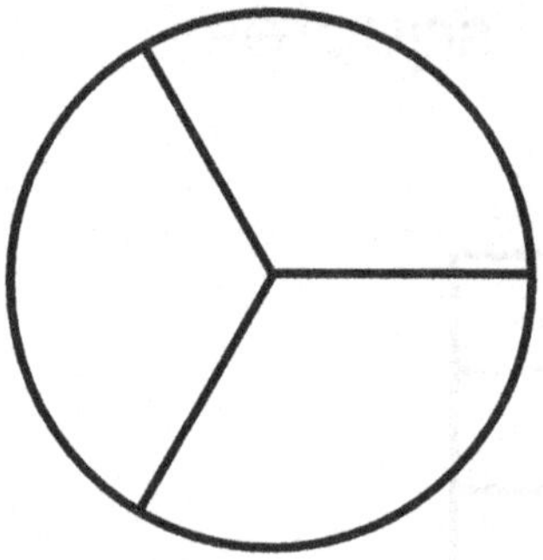

halves

thirds

quarters

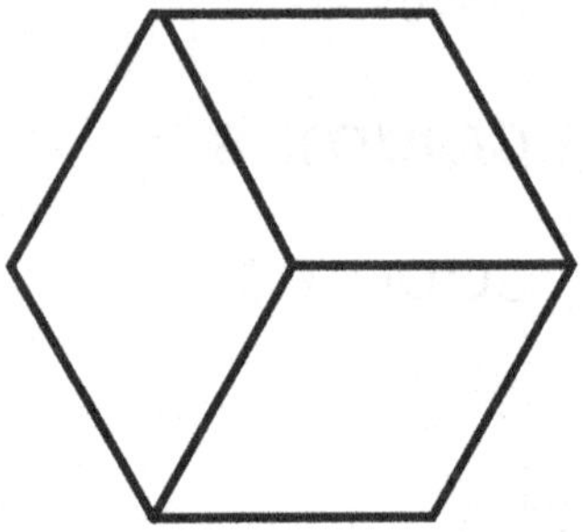

halves

thirds

quarters

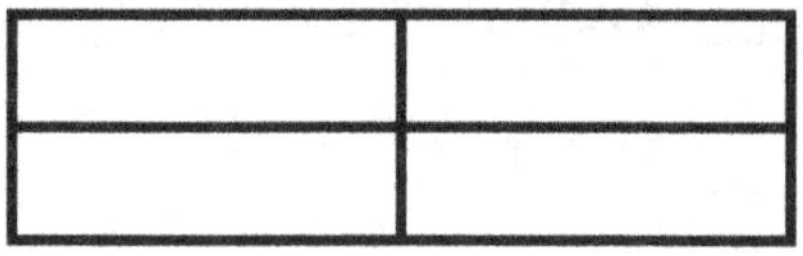

halves

thirds

quarters

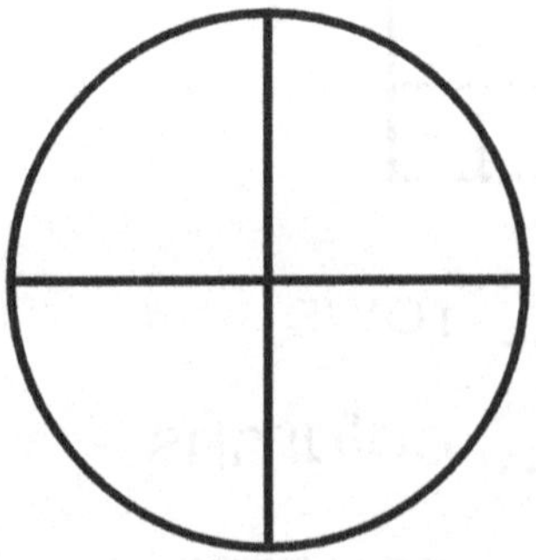

halves

thirds

quarters

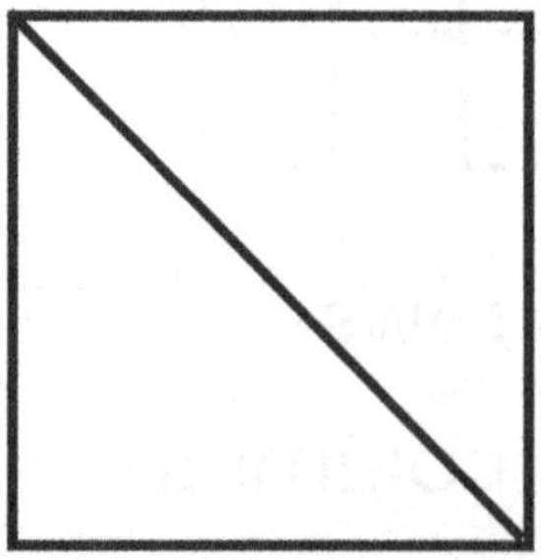

halves

thirds

quarters

# Shapes to Represent Fractions



Write whether each shape is divided into **halves**, **thirds**, or **fourths**.

EX

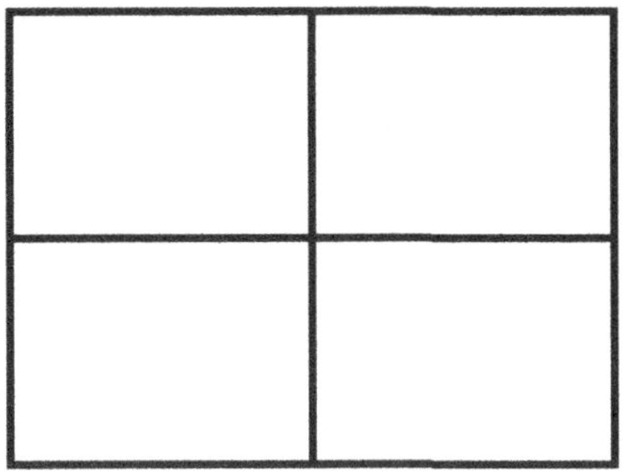

fourths

1

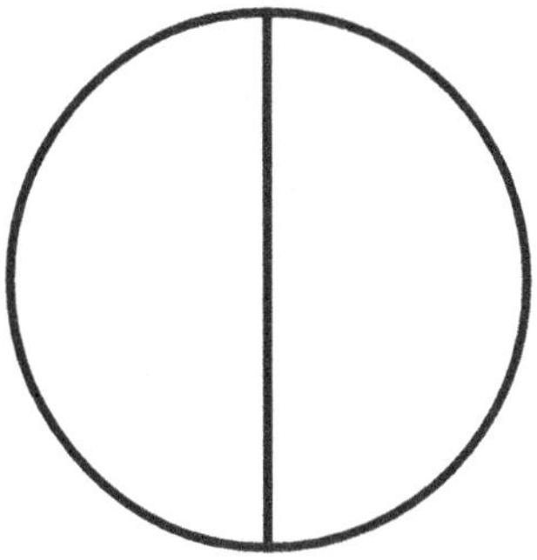

2

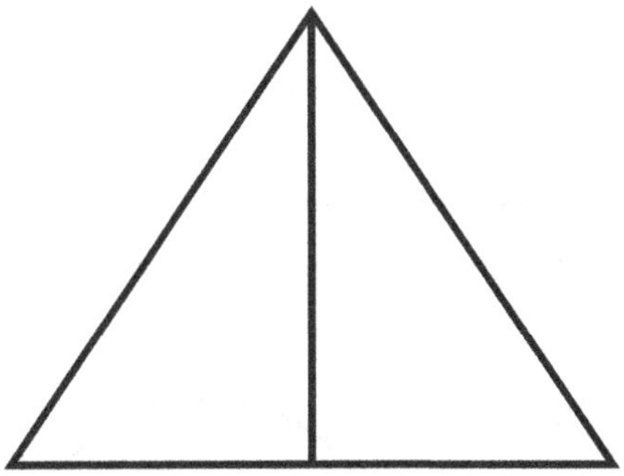

3

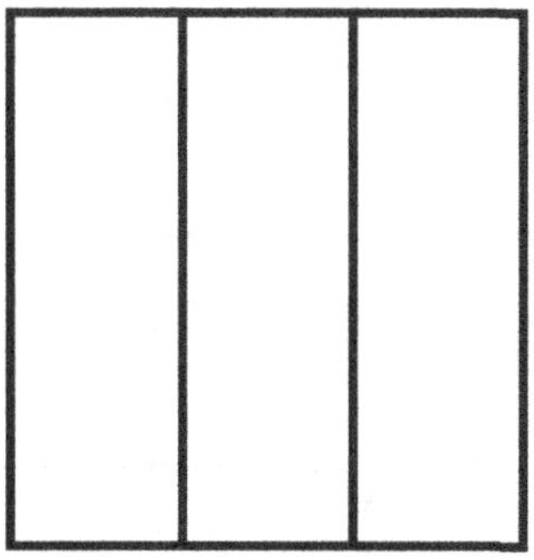

4

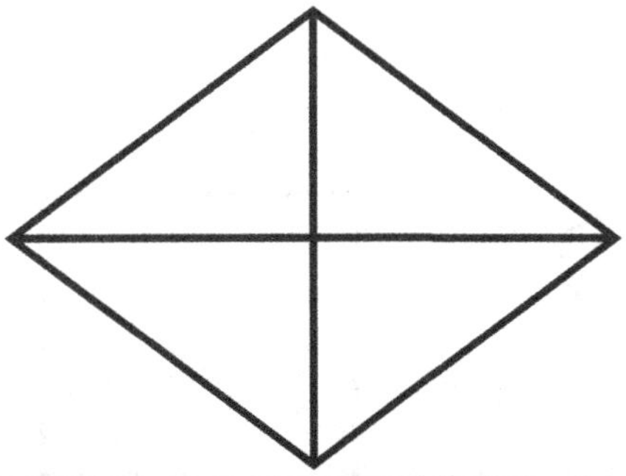

5

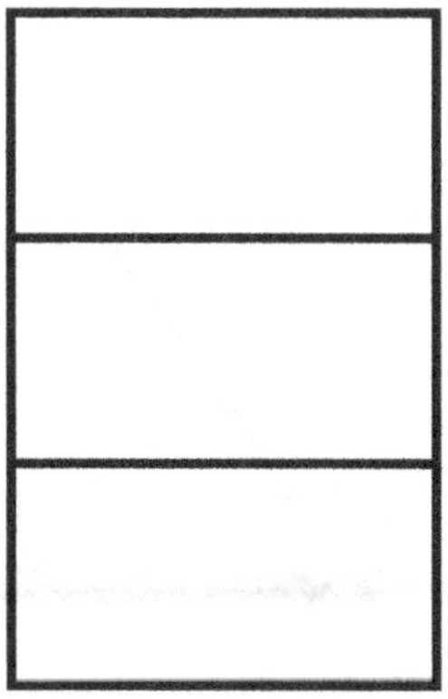

# Shapes to Represent Fractions

Draw lines to divide each shape into halves. Color one half.

1
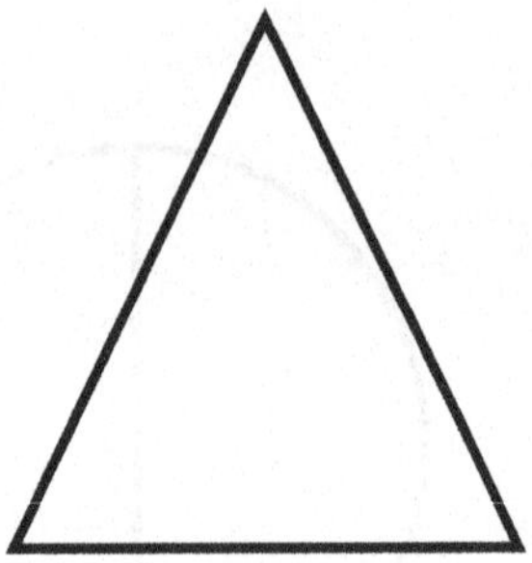

2
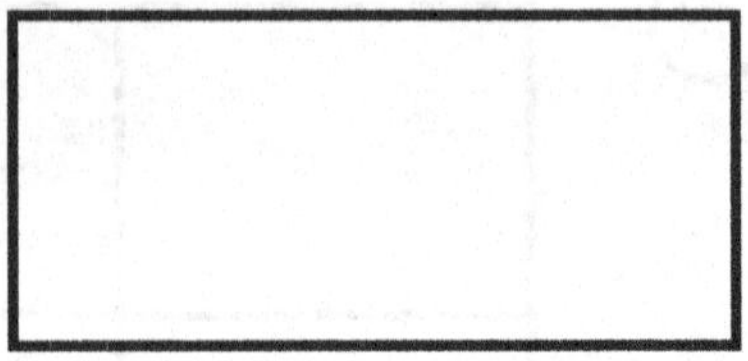

3
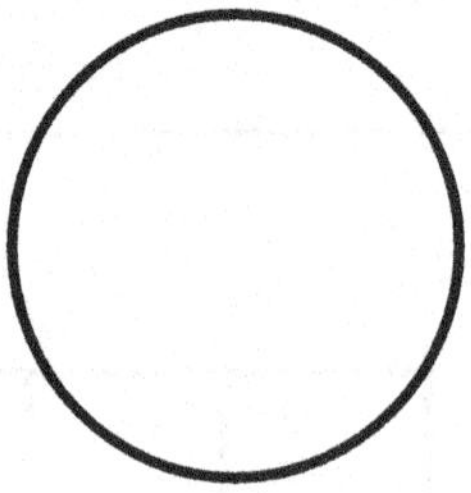

4
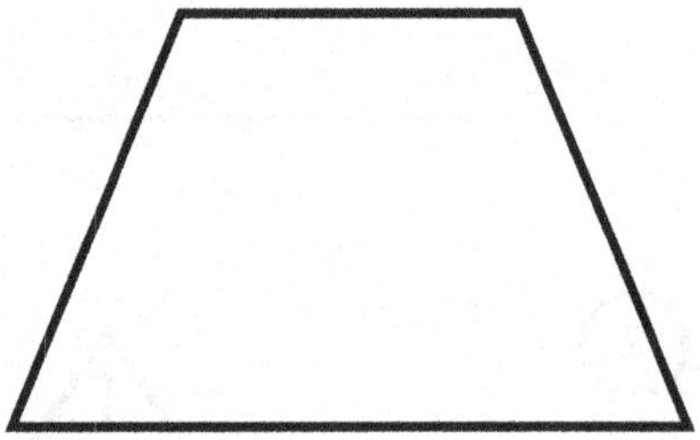

Draw lines to divide each shape into quarters. Color one fourth.

1

2
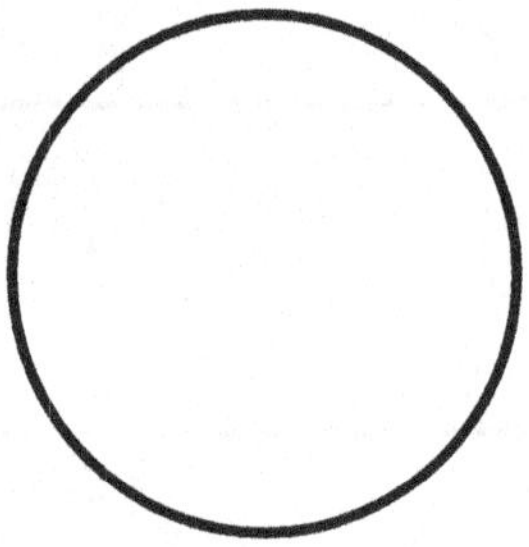

3
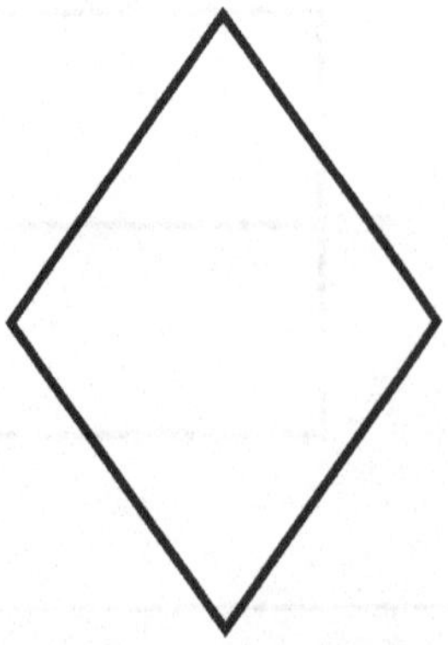

4
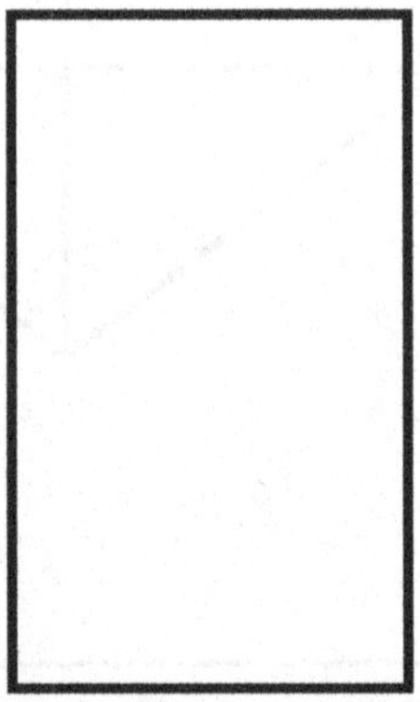

Color one share of each shape. Write **half**, **third**, or **fourth** to complete each sentence.

(EX)

One ___third___ of the square is shaded.

(1)

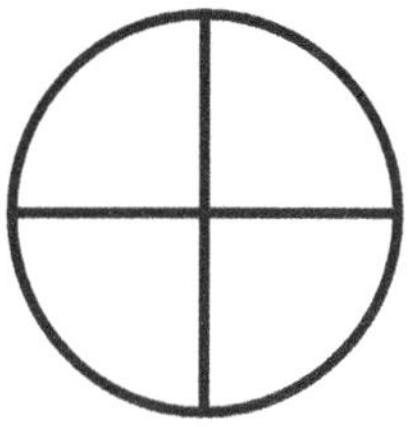

One __________ of the circle is shaded.

(2)

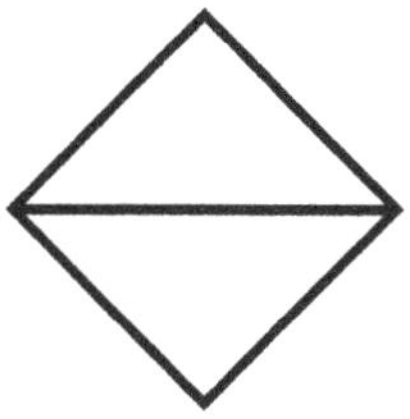

One __________ of the diamond is shaded.

(3)

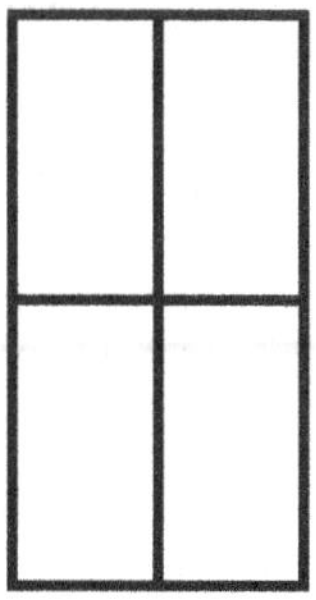

One __________ of the rectangle is shaded.

(4)

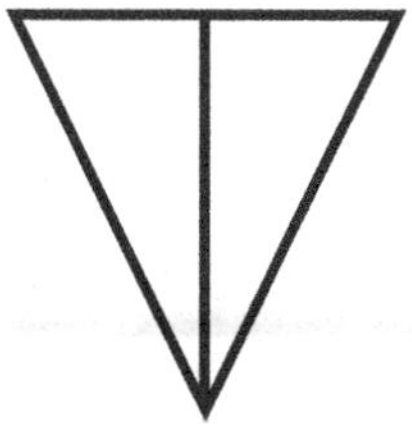

One __________ of the triangle is shaded.

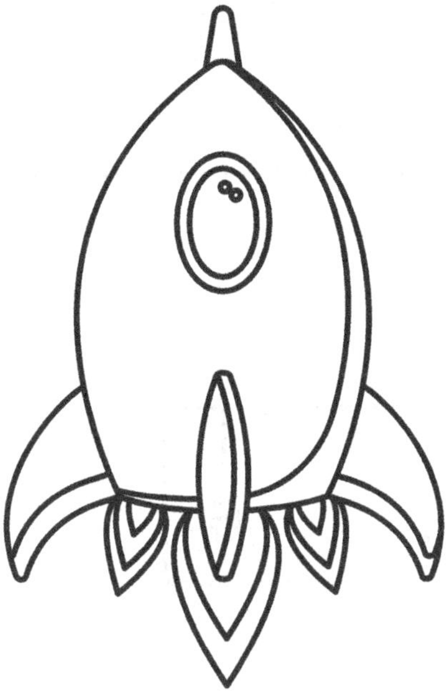

# Measurement and Data

# Measuring Length

Use the inch ruler shown to estimate the length of each object in inches.

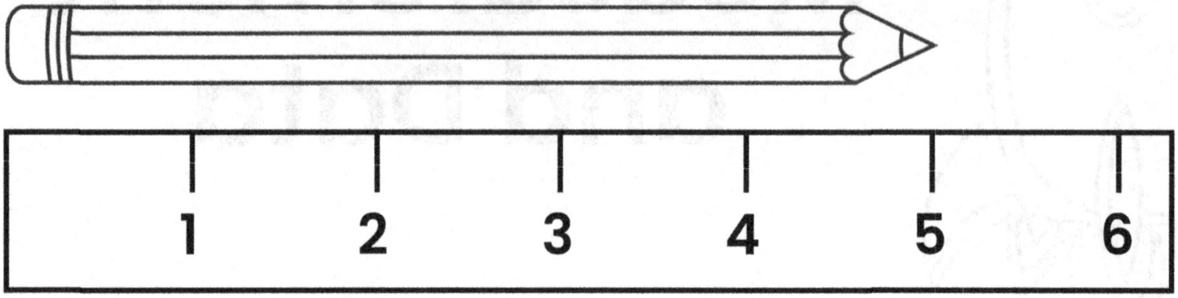


______ inches

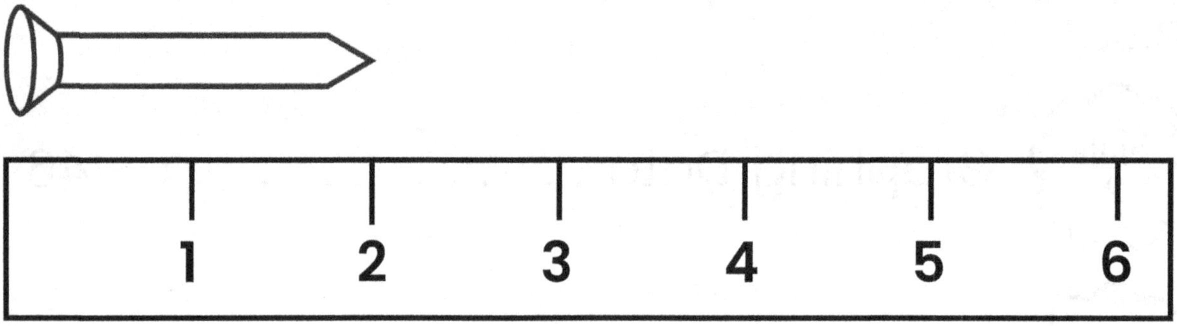


______ inches

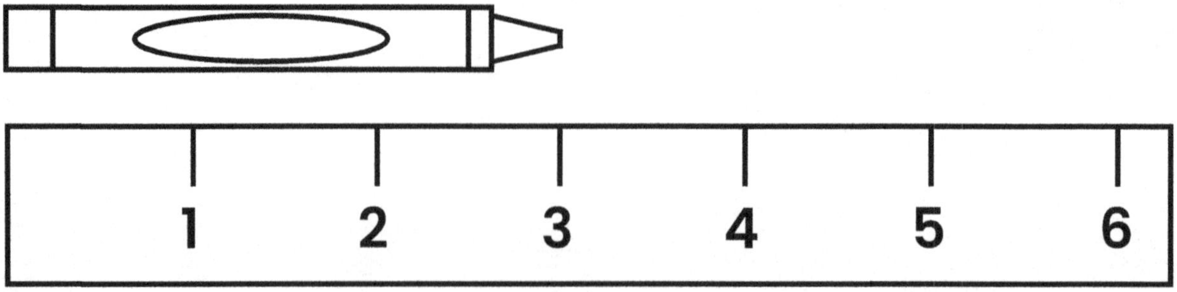


______ inches

# Measuring Length



Use a ruler or measuring tape to measure each item to the nearest inch.

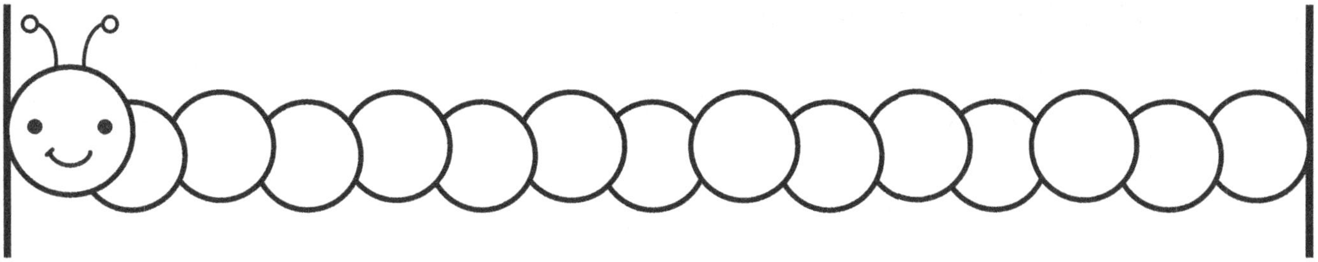

_____ inches

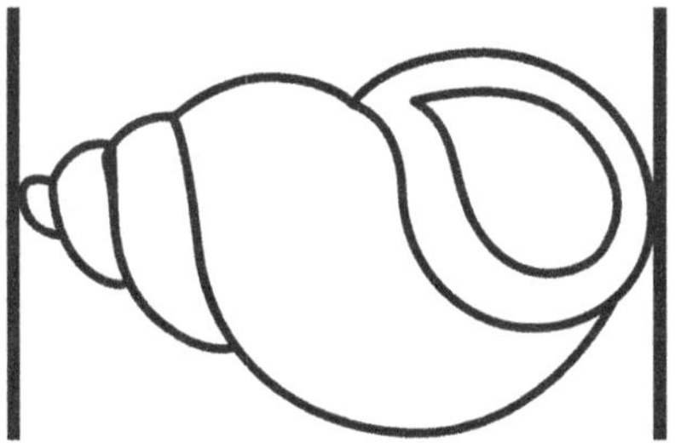

_____ inches

_____ inches

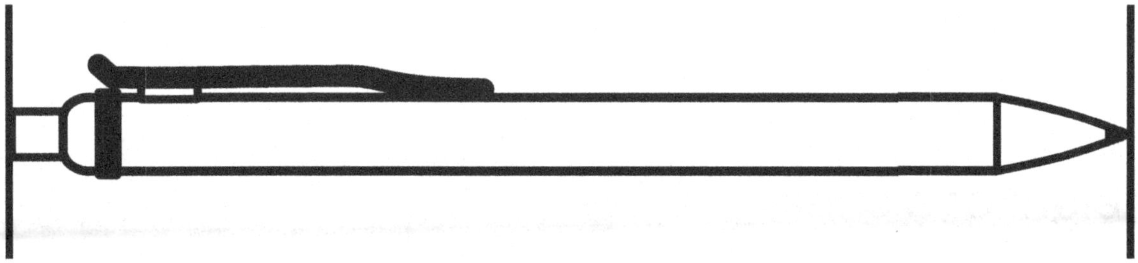

_____ inches

# Measuring Length

Use a ruler or measuring tape to find the lengths of these real-life objects in both inches and feet (to the nearest foot).

1. The width of a window

   _______ inches _______ feet

2. The length of a desk or table

   _______ inches _______ feet

3. The height of a person

   _______ inches _______ feet

4. The length of a poster, banner, or other picture

   _______ inches _______ feet

5. The height of a chair

   _______ inches _______ feet

Use a yardstick to find the lengths of these real-life objects to the nearest yard.

1. A vehicle

_______ yards

2. The room you are in

_______ yards

3. A countertop

_______ yards

4. A long piece of furniture like a sofa, bench, or bed

_______ yards

5. A large item of your choice

_______ yards

# Measuring Length

Fill in the missing numbers to show the same lengths in inches, feet, and yards as each length given.

| Inches | Feet |
|---|---|
| 12 | |

| Inches | Feet |
|---|---|
| 48 | |

| Inches | Feet |
|---|---|
| | 2 |

| Inches | Feet |
|---|---|
| | 5 |

| Inches | Feet | Yards |
|---|---|---|
| 36 | | |

| Inches | Feet | Yards |
|---|---|---|
| | | 2 |

Fill the the correct bubble to answer each question.

Which of these is most likely to be the length of a notebook?

○ 10 inches ○ 10 feet ○ 10 yards

Which of these is most likely to be the length of a blanket?

○ 6 inches ○ 6 feet ○ 6 yards

Which of these is most likely to be the length of a motorcycle?

○ 2 inches ○ 2 feet ○ 2 yards

Which of these is most likely to be the length of a pen?

○ 2 inches ○ 7 inches ○ 18 inches

Which of these is most likely to be the length of a bicycle?

○ 1 foot ○ 5 feet ○ 10 feet

# Measuring Length

Use the centimeter ruler shown to estimate the length of each object in centimeters.

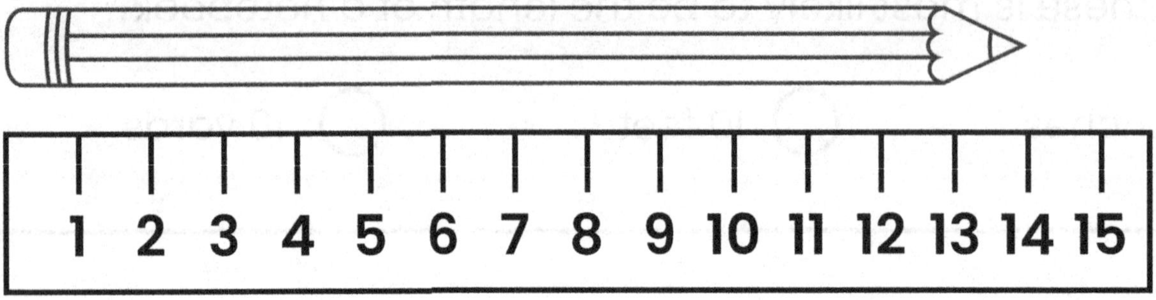


______ centimeters

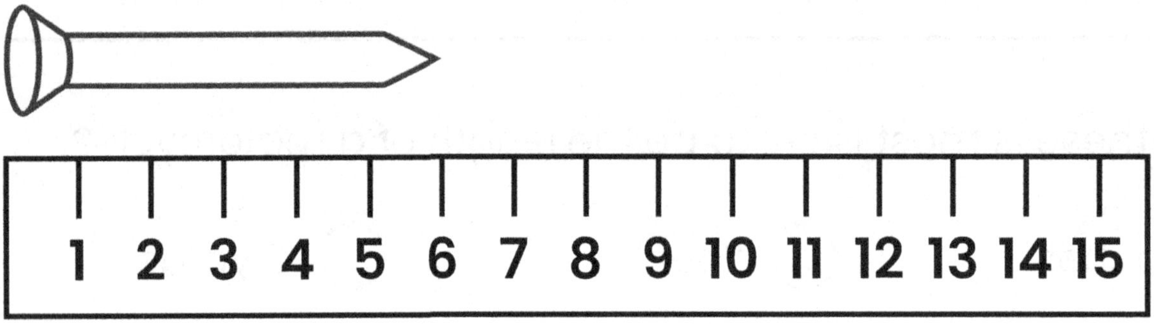


______ centimeters

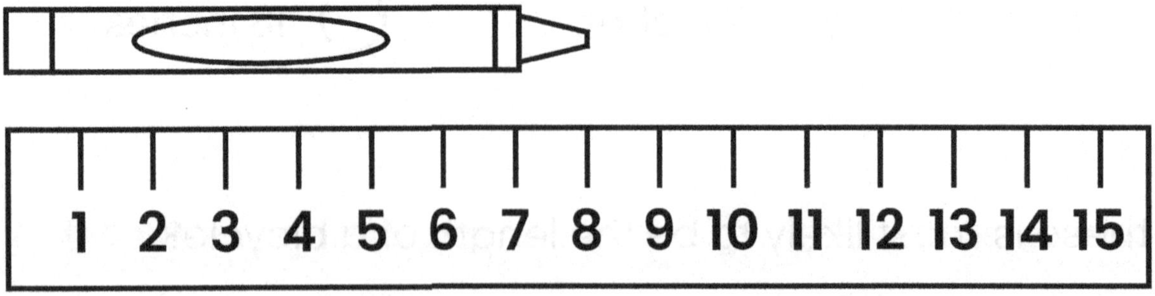


______ centimeters

# Measuring Length



Use a ruler or measuring tape to measure each item to the nearest centimeter.

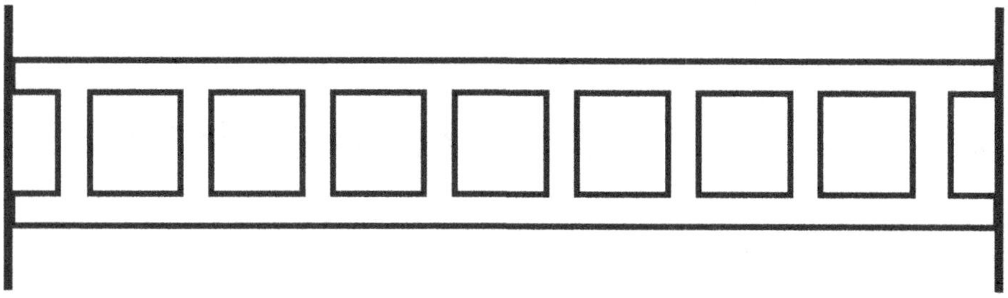

______ centimeters

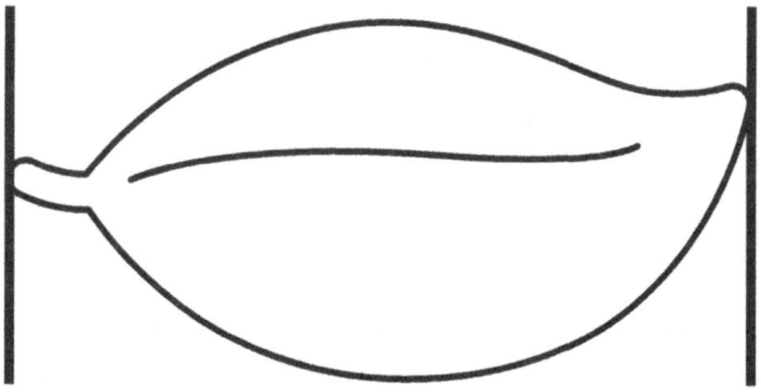

______ centimeters

______ centimeters

______ centimeters

# Measuring Length

Use a ruler or measuring tape to find the lengths of these real-life objects in centimeters.

1. A paperclip

   _______ centimeters

2. A marker

   _______ centimeters

3. A piece of paper

   _______ centimeters

4. Your hand

   _______ centimeters

---

Use a meterstick to find the lengths of these real-life objects to the nearest meter.

1. The height of a person

   _______ meters

2. A rug

   _______ meters

3. A bookcase

   _______ meters

4. A jump rope

   _______ meters

# Measuring Length



Fill the the correct bubble to answer each question.

Which of these is most likely to be the length of a picture frame?

○ 2 centimeters ○ 2 meters ○ 20 centimeters

Which of these is most likely to be the length of a phone?

○ 1 centimeter ○ 14 centimeters ○ 40 centimeters

Which of these is most likely to be the length of a car?

○ 5 centimeters ○ 50 centimeters ○ 500 centimeters

Which of these is most likely to be the length of a dining table?

○ 2 meters ○ 6 meters ○ 10 meters

Which of these is most likely to be the length of a hot dog?

○ 2 centimeters ○ 15 centimeters ○ 1 meter

# Measuring Length

Answer the questions based on the pictures below showing a rope, a crayon, and a centimeter ruler.

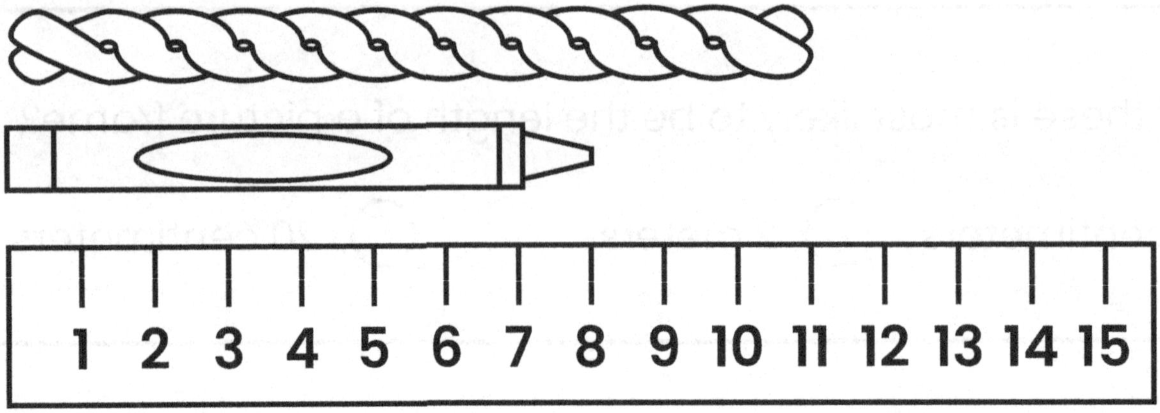


1. How long is the rope? ______ centimeters

2. How long is the crayon? ______ centimeters

3. How much longer is the rope than the crayon?

   ______ centimeters

---

Fill in the missing numbers to show the same lengths in centimeters and meters as each length given.

| Centimeters | Meters |
|---|---|
| 100 | |

| Centimeters | Meters |
|---|---|
| | 3 |

| Centimeters | Meters |
|---|---|
| | 4 |

| Centimeters | Meters |
|---|---|
| 200 | |

Answer the word problems about measurement.

Rosa is 52 inches tall. Alan is 48 inches tall. How much taller is Rosa than Alan?

_____ inches

Kelsey runs for 70 meters. Hannah runs for 110 meters. How far do they run all together?

_____ meters

Terry uses a measuring tape to find the distance around 2 balls. The soccer ball is 68 centimeters and the baseball is 23 centimeters. How much larger is the soccer ball?

_____ centimeters

Use the picture graph to answer the questions below.

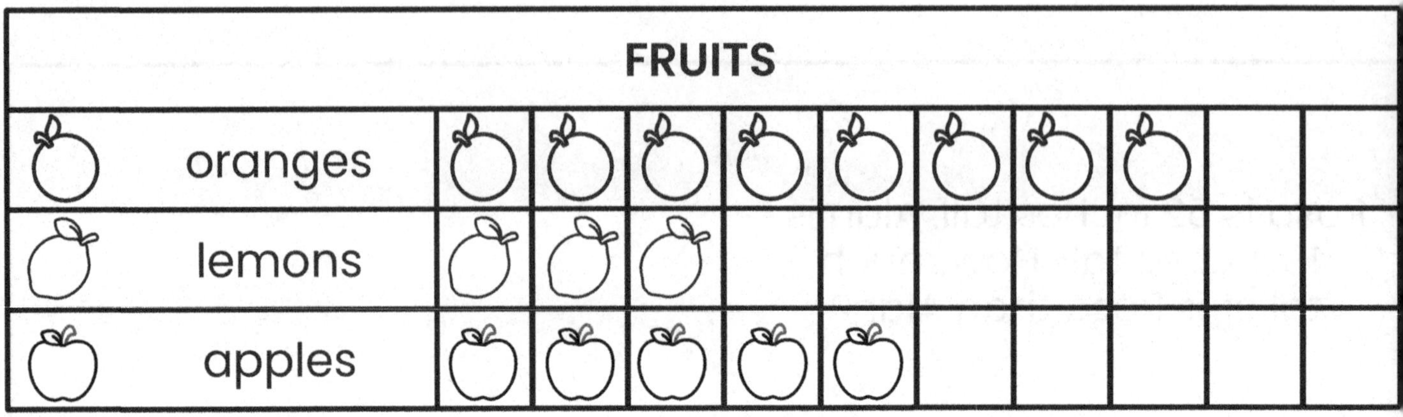


① How many of each fruit are there?

 ______  ______  ______

② How many pieces of fruits are there all together?

______ + ______ + ______ = ______

③ How many more oranges are there than apples?

______ - ______ = ______

④ How many fewer lemons are there than oranges?

______ - ______ = ______

Fill in the totals on the tally chart. Then, use the tally chart to answer the questions below.

| STUDENTS' FAVORITE ICE CREAM FLAVORS | | Total |
|---|---|---|
| chocolate |  | |
| vanilla |  | |
| strawberry |  | |
| peanut butter |  | |

1. Which flavor was chosen by the most students? ______________

2. How many more students chose chocolate than peanut butter? ______

3. How many students chose either vanilla or strawberry? ______

---

Complete the tally chart based on the picture below.

| INSECTS | | Total |
|---|---|---|
|  | | |
|  | | |
|  | | |

Use the bar graph to answer the questions below.

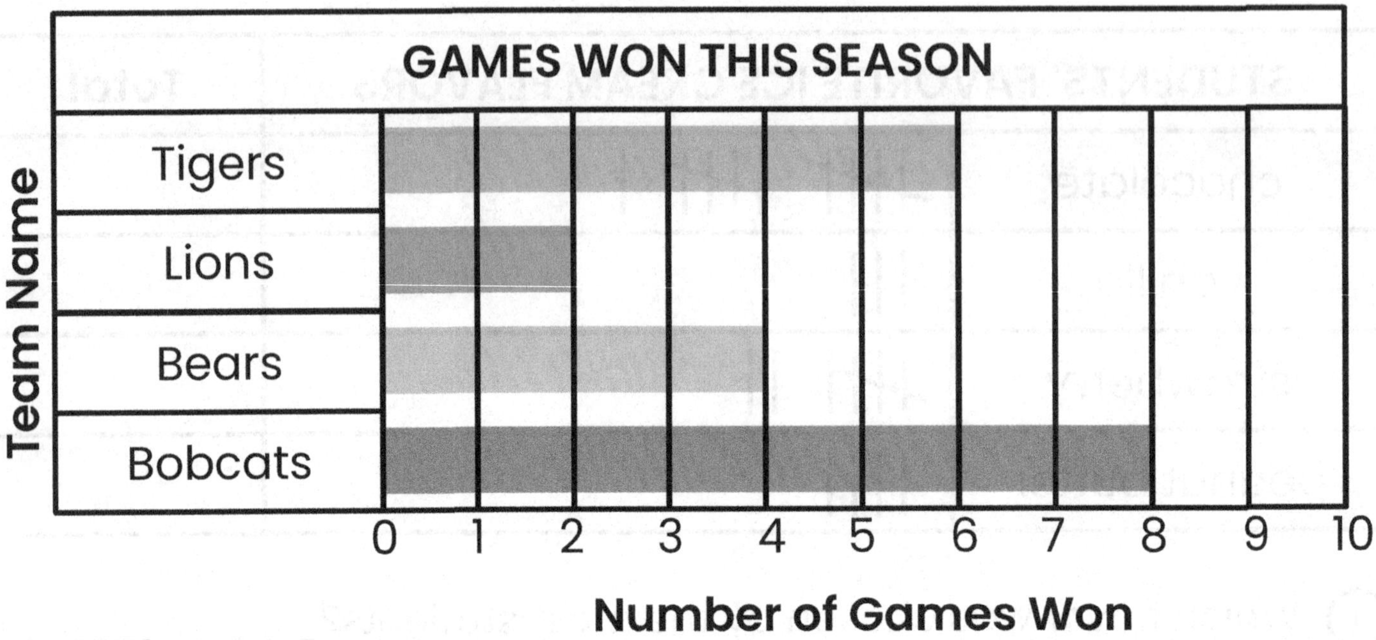


1. How many games did the Bears win? ______

2. How many games did the Lions and the Tigers win? ______

3. How many more games did the Tigers win than the Bears? ______

4. How many more games did the Bobcats win than the Lions? ______

Make a bar graph to show the data given. Then, answer the questions below.

Mr. Tate asked the students in his class what their favorite season was. 4 students chose spring, 9 students chose summer, 5 students chose fall, and 3 students chose winter.

**STUDENTS' FAVORITE SEASON**

| Season | | | | | | | | | | |
|---|---|---|---|---|---|---|---|---|---|---|
| spring | | | | | | | | | | |
| summer | | | | | | | | | | |
| fall | | | | | | | | | | |
| winter | | | | | | | | | | |
| | 0 | 1 | 2 | 3 | 4 | 5 | 6 | 7 | 8 | 9 | 10 |

**Number of Students**

1. Which season is the least popular? ______________

2. How many students chose spring or fall? ______

3. How many more students chose summer than fall? ______

4. How many students are in the class? ______

# 150 Graphing Data

Make a bar graph to show the data given.

Students chose their favorite animal that they saw on a field trip to the zoo.

- 5 students chose elephants.
- 7 students chose penguins.
- 3 students chose giraffes.
- 4 students chose tigers.

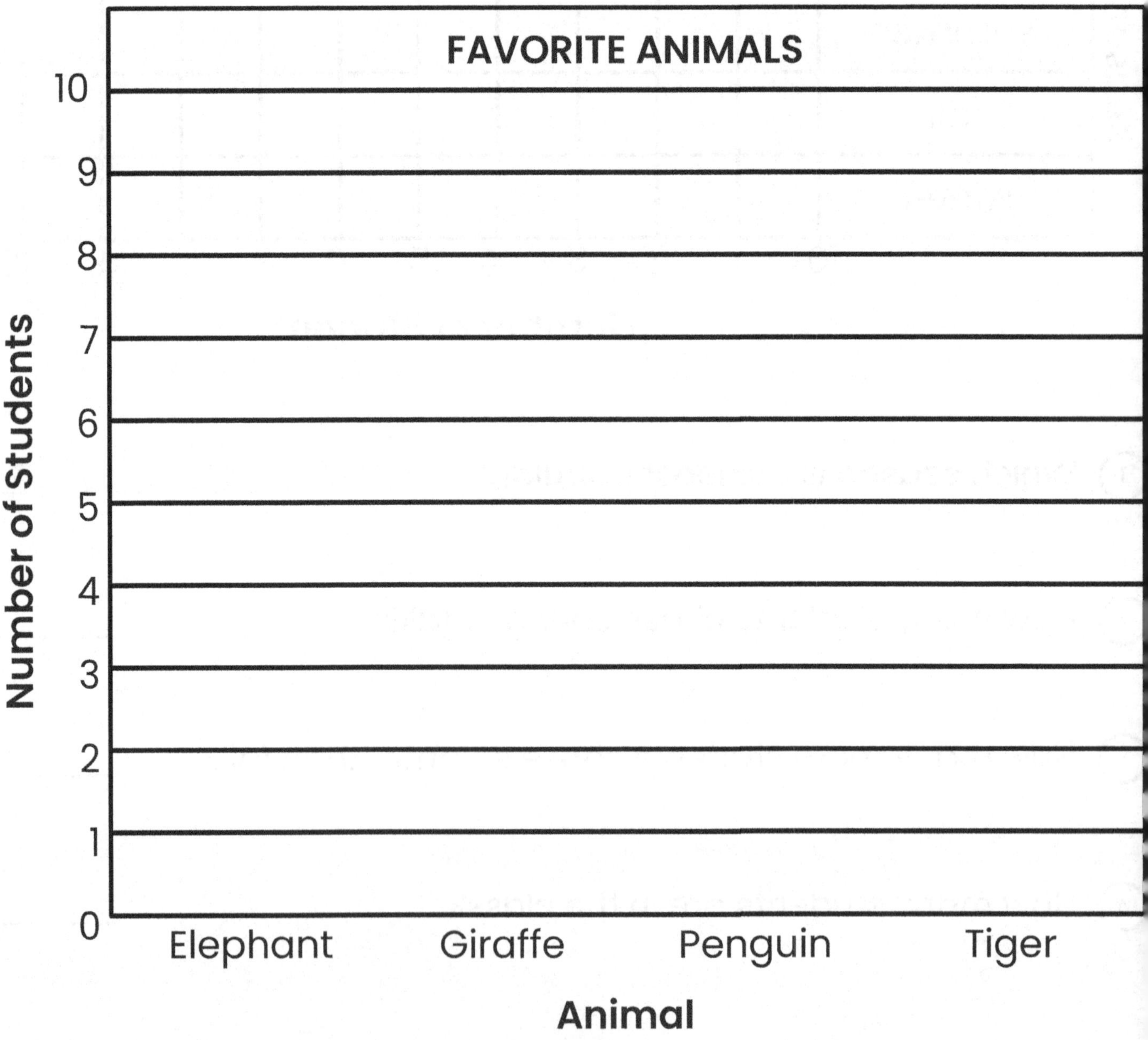

Answer the question below based on the graph you made on the previous page.

1. How many students chose penguins or giraffes as their favorite animal? ______

2. How many more students chose penguins than tigers? ______

3. How many students chose tigers or elephants as their favorite animal? ______

4. How many students are there all together? ______

5. How many fewer students chose tigers than elephants? ______

6. How many more votes did the most popular animal get than the least popular animal? ______

# Time

Draw hands onto each clock to show the correct time.

| | | | |
|---|---|---|---|
|  | 10:15 |  | 5:40 |
|  | 2:30 |  | 11:05 |
|  | 3:35 |  | 9:50 |
|  | 6:10 |  | 1:25 |

Write the time shown on each clock.

For each time given, draw hands on the analog clock on the left and numbers on the digital clock on the right.

eleven o'clock

half past ten

a quarter past one

five minutes past noon

Fill in the correct bubble to answer each question.

Which time is the same as a quarter past four in the afternoon?

- ○ 4: 15 a.m.
- ○ 4:15 p.m.
- ○ 4:30 p.m.

Which time shows midnight?

- ○ 12:00 a.m.
- ○ 6:00 p.m.
- ○ 12:00 p.m.

Which time is the same as eight thirty in the morning?

- ○ 8:00 a.m.
- ○ 8:30 a.m.
- ○ 8:30 p.m.

Which of these is most likely to be the time someone eats lunch?

- ○ 1:00 a.m.
- ○ 11:30 a.m.
- ○ 12:30 a.m.

Which of these is most likely to be someone's bedtime?

- ○ 7:15 a.m.
- ○ 11:00 a.m.
- ○ 9:00 p.m.

# Money

Write the value of each coin.

_______ ¢

_______ ¢

_______ ¢

_______ ¢

_______ ¢

_______ ¢

_______ ¢

_______ ¢

---

Write the total value of the coins below.

_______ ¢

Write the name of each coin on the line.

______________________

______________________

______________________

______________________

---

Write the total value of the coins.

1. 1 quarter and 2 dimes ________ ¢

2. 1 dime, 1 nickel, and 3 pennies ________ ¢

3. 1 quarter, 1 dime, 2 nickels, and 4 pennies ________ ¢

# Money

Fill in the blanks with a number to make each sentence true.

1. 1 dime is equal to ______ nickels.

2. 2 quarters are equal to ______ dimes.

3. 4 dimes are equal to 1 quarter and ______ pennies.

4. 5 dimes and 5 nickels are equal to ______ quarters.

---

Allie and Shane each have the coins shown below. Fill in the blanks to tell about their coins.

Allie's Coins

Shane's Coins

1. Allie has ______ ¢ and Shane has ______ ¢.

2. Together, Allie and Shane have ______ ¢.

3. Allie has ______ ¢ more than Shane.

Write or draw two different ways to use coins to make the amount shown.

28¢

17¢

41¢

60¢

# 160 Money

Fill in the blanks with the correct number to make each statement true.

① $1.00 = ________ pennies

② $1.00 = ________ nickels

③ $1.00 = ________ dimes

④ $1.00 = ________ quarters

---

Tom has the coins shown below. Answer the questions that follow.

① How much money does Tom have? ____________________

② How much more would Tom need to make $1.00? ____________

③ Which coins could make that amount of money? ____________

______________________________________________

Count the money. Write the total on the line.

$ ______

$ ______

$ ______

$ ______

# Money

Answer the word problems about money in a complete sentence.

Landon has two $10 bills, two $5 bills, and four $1 bills. How much money does he have?

______________________________

______________________________

Abby has $12. She gets $5 each week for allowance. How much money will she have at the end of two weeks?

______________________________

______________________________

Ashley has three quarters. Gina has 7 dimes. How much more money does Ashley have than Gina?

______________________________

______________________________

Answer the word problems about money in a complete sentence.

Keenan has one $20 bill, two $5 bills, and three $1 bills. If he spends $30 on a video game, how much money will he have left?

______________________________

______________________________

Omar has $1. He buys candy that costs 65¢. How much change will he get back?

______________________________

______________________________

A cheeseburger costs $2. How many cheeseburgers could you buy with a $10 bill?

______________________________

______________________________

Made in United States
North Haven, CT
10 November 2026

22106119R00089